Sand Talk

*How Indigenous Thinking
Can Save the World*

o o o

Tyson Yunkaporta

HarperOne
An Imprint of HarperCollinsPublishers

HarperOne

HarperCollins books may be purchased for educational, business, or sales promotional use. For information, please email the Special Markets Department at SPsales@harpercollins.com.

Originally published as *Sand Talk* in Australia in 2019 by The Text Publishing Company.

First HarperOne hardcover published in 2020.

FIRST EDITION

Images courtesy of the author, copyright © 2020.

Designed by Ad Librum

Library of Congress Cataloging-in-Publication Data
Names: Yunkaporta, Tyson, author.
Title: Sand talk : how indigenous thinking can save the world / Tyson Yunkaporta.
Description: First edition. | San Francisco, California : HarperOne, 2020 | Originally published as 'Sand Talk' in Australia in 2019 by The Text Publishing Company.
Identifiers: LCCN 2019050676 (print) | LCCN 2019050677 (ebook) | ISBN 9780062975645 (hardcover) | ISBN 9780062975638 (ebook)
Subjects: LCSH: Civilization, Modern—21st century—Philosophy. | Sustainability—Philosophy. | Philosophy, Aboriginal Australian.
Classification: LCC CB430 .Y86 2020 (print) | LCC CB430 (ebook) | DDC 909.82—dc23
LC record available at https://lccn.loc.gov/2019050676
LC ebook record available at https://lccn.loc.gov/2019050677

20 21 22 23 24 LSC 10 9 8 7 6 5 4 3 2 1

Sand Talk

The Porcupine, the Paleo-mind, and the Grand Design

Sometimes I wonder if echidnas ever suffer from the same delusion that many humans do, that their species is the intelligent center of the universe. These Australian porcupines are smart enough: their prefrontal cortex, the area of the brain used for complex reasoning and decision-making, is the biggest of any mammal in relation to body size. Fifty percent of the echidna brain is used for some of the hardest kinds of thinking. In humans, it is not even 30 percent.

In acknowledging this, I am paying my respects to the sentient totemic entities all over Australia where these echidnas follow the songlines of their creation: maps of story carrying knowledge along the lines of energy that manifest as Law in the mind and land as one, webbed throughout the traditional lands of the First Peoples.

You might join me in paying respects to the people and other beings everywhere who keep the Law of the land:

The Elders and traditional custodians of all the places where this book is written and read.

The Ancestors, the old people from every People now living on the continent currently known as Australia and its islands.

Our nonhuman kin, including the various spiky species around the world, the porcupines and hedgehogs who snuffle in the earth for ants and then do God knows what when we're not looking.

I don't know why Stephen Hawking and others have worried about superintelligent beings from other planets coming here and using their advanced knowledge to do to the world what industrial civilization has already done. Beings of higher intelligence are already here, always have been. They just haven't used their intelligence to destroy anything yet. Maybe they will, if they tire of the incompetence of domesticated humans.

All humans evolved within complex, land-based cultures over deep time to develop a brain with the capacity for over one hundred trillion neural connections, of which we now use only a tiny fraction. Most of us have been displaced from those cultures of origin, a global diaspora of refugees severed not only from land but from the sheer genius that comes from belonging in symbiotic relation to it. In Aboriginal Australia, our Elders tell us stories, ancient narratives to show us that *if you don't move with the land, the land will move you.* There is nothing permanent about settlements and the civilizations that spawn them. Maybe the reason all the powerful instruments pointed at the sky have not yet been able to detect high-tech alien civilizations is that these unsustainable societies don't last long enough to leave a cosmic trace. An unsettling thought.

Perhaps we need to revisit the brilliant thought-paths of our Paleolithic Ancestors and recover enough cognitive function to correct the impossible messes civilization has created, before the echidnas decide to sack us all and take over as the custodial species of this planet.

The stories that define our thinking today describe an eternal battle between good and evil springing from an originating act of sin. But these terms are just metaphors for something more difficult to explain, a relatively recent demand that simplicity and order be imposed upon the complexity of creation, a demand sprouting from an ancient seed of narcissism that has flourished due to a new imbalance in human societies.

There is a pattern to the universe and everything in it, and there are knowledge systems and traditions that follow this pattern to maintain balance, to keep the temptations of narcissism in check. But recent traditions have emerged that break down creation systems like a virus, infecting complex patterns with artificial simplicity, exercising a civilizing control over what some see as chaos. The Sumerians started it. The Romans perfected it. The Anglosphere inherited it. The world is now mired in it.

The war between good and evil is in reality an imposition of stupidity and simplicity over wisdom and complexity.

○

A collection of pages filled with marks representing speech sounds is a complicated way of communicating, particularly when you want to convey a practical sense of the pattern of creation that might shed light on current crises the world is facing. Complicated, not complex. They are two very different things. Viewing the world through a lens of simplicity always seems to make things more complicated but simultaneously less complex.

For an Indigenous Australian coming from an intensely interdependent and interpersonal oral culture, writing speech-sound symbols for strangers to read makes things even more complicated. That is exacerbated when the audience is preoccupied with notions of authenticity and the writer's standing as a member of a cultural minority that has lost the right to define itself. The ability to write fluently in the language of the occupying power seems to contradict an Indigenous author's membership in a community that is not supposed to be able to write about itself at all. So at this point I will need to explain who I am and how I came to be writing this.

In my own world I know myself as my community knows me: a boy who belongs to the Apalech clan from Western Cape York, a Wik Mungkan speaker with ties to many language groups on the continent currently called Australia, including adoptive ties. Some adoptive ties are informal, such as those I have in New South Wales and Western Australia, but my customary adoption two decades ago into Apalech is under Aboriginal Law, which is strict and inalienable. This Law prevents me from identifying with Nungar/Koori/Scottish affiliations by descent and demands that I take on exclusively the names and roles and genealogies required of Apalech clan membership. I honor this no matter what, even though I know most people don't understand it, and it makes me look silly: while people in the south tell me I look Indian or Aboriginal or Arab or Latino, when I stand beside my very dark-skinned adoptive father, I look like Nicole Kidman.

My life story is not redemptive or inspiring in any way, and I don't like sharing it. It shames and traumatizes me, and I need to protect myself as well as others who have been thrown about in the cyclones of this messy colonial history. But people insist on knowing about it before reading my work, for some reason, so here is the condensed version.

I was born in Melbourne but relocated north as an infant, then grew up in a dozen different remote or rural communities all over Queensland, from Benaraby to Mount Isa. After a challenging and often horrific period of schooling, I was eventually unleashed on the world as an angry young male, in a flurry of flying fists and cultural dysphoria. Combine the worst parts of the films *Once Were Warriors*, *Conan the Barbarian*, and *Goodfellas*, and you'll get a fair idea of what went on. As a child I was not a happy camper, but taking control of my life as a legal adult did not improve my disposition, and for that I blame nobody but myself.

Finding and reconnecting with my "tribe" down south did not live up to the homecoming fantasy I had imagined for so long, and this left me feeling quite devastated and alone. But it wasn't all bad. I was lucky enough to pick up a lot of fragmentary land-based and cultural knowledge on my life's journey up to that point. In the 1990s I worked as a teacher, running Aboriginal student-support programs in schools, teaching drama and languages, making my didgeridoos and spears and clapsticks, and dancing corroboree and hunting kangaroos and performing the exotica of my culture that I'd learned over the years. But it was all disconnected and hollow, just fragments and window dressing. I cringe when I think about it.

Although in the middle of all this mess I somehow managed to study, get married, and have two beautiful children, my life had been so defined by patterns of violence and substance abuse that I was not even a real person—just a bundle of extreme reactions and rage. In my late twenties I found myself in the far north, a rogue without family or purpose. I had lived too long with the label "part-Aboriginal" or "touch of the tar" and was ridiculed for it in the institutions where I worked or studied. I wasn't coping well with the endless cycles of interrogation about my identity. "You're not white. What nationality are you? Aboriginal? Nah, you look white. What percent Aboriginal are you? Well, we've all got a bit in us. Most white Australians could get an Aboriginality certificate if they did a blood test and a family tree."

Up north, the racist abuse I encountered pushed me over the edge. I went off the rails completely, and it was nearly the end of me. One terrible night Dad Kenlock and Mum Hersie found me in a moment of self-inflicted peril and saved my life. They had lost their youngest son the year before—he was my age when he met with the same peril—and they decided to raise me as their own. I've belonged to them ever since.

So this family became the center of my life, and I orbited around it, living longer on Cape York than in any other place I'd lived before, and taking family members with me to stay down south when I went away to work in different temporary jobs. This gave them access to quality education and services that were not available in our home community. There was no substantial work available there either, so Dad Kenlock told me to go out and use my knowledge to "fight for Aboriginal rights and culture."

I traveled out periodically from my home base to work with Indigenous groups and communities all over Australia, while my own poor kids and their mother, and my extended family, endured my long absences. I gained more knowledge, but it was at a price. I needed to work and study hard so I could support my children and extended family dependents, but I also needed to live and grow in my culture. Those are big things. Nobody can do both without damaging their most important relationships. The attempt eventually cost me my marriage. I missed a lot of funerals and birthdays and became a cautionary tale in my community: "Too much work and education, no good. You finish up like brother Ty."

But what I gained was important. I lived out in the bush for much of this period and formed close bonds with a lot of Elders and knowledge-keepers across Australia, and they taught me more about the old Law, the Law of the land. I worked with Aboriginal languages, schools, ecosystems, research projects, wood carvings, philanthropic groups, and songlines.

In my travels I saw that it was our ways, not our things, that grounded us and sustained us. So I began to find words and images to express those Indigenous patterns of thinking, being, and doing that are usually invisible and obscured by a focus on exotic items and performances. I started translating those ideas into English print so others could understand them and so our own people could assert them, completing master's and doctoral degrees and publishing papers as I went. I started writing articles from this point of view when I moved to Melbourne, spending some time living and working in my place of birth. I was asked to write a book about the articles I wrote in that period, and so here we are. I'm writing this just down the road from the place where I was born, while struggling to adjust to city life and clean up the messes I've made over the last five decades.

Like I said, this is not an inspirational tale of redemption or triumph over adversity. I'm not a success story or role model or expert or anything like that. I am still a reactive and abrasive boy who is terrified of the world, although this is moderated now by a core of calm and intelligence my family has worked hard to develop in me. This is the thing that keeps me breathing, along with a network of relations and cultural affiliations all over the continent that I have obligations to, demanding I move in the world with respect and care. Or try to: I don't always succeed. But there are many people who care for me and defend me no matter what, and when I travel around there is always a bed, a yarn, and a feed waiting for me. My woman, my children, and my community hold me up and watch my back, as I watch theirs. I know who I am, where I belong, and what I call myself, and it is enough.

When I'm away from my community, though, there are people who want to sort me into unfamiliar categories, and I often don't get to decide what to call myself. I frequently have to call myself *Bama* because culturally senior people in the south have insisted on it. Never mind that I know the word just means "man," and I say it with a "p" rather than a "b." Or that in my community the only cultural situation where a person would actually call themselves *pama* is if they were looking to start a fight by proclaiming their exceptional manhood: "*Ngay pama!* I'm a man!" Or that, in fact, I'm un-initiated, which means that at the age of forty-seven I still only have the cultural knowledge and status of a fourteen-year-old boy. A swimming pool was built on the initiation ground back home, so those rites of passage don't happen anymore. But when in Rome I try to do as the Romans do, so *Bama* it is in most introductions requiring me to break my identity into digestible chunks.

○

Speaking of Rome, it must be acknowledged that there is nothing new about imperial cultures imposing classifica-tions on Indigenous people. The Romans classified the Gauls in three groups this way: the toga-wearing Gauls (basically, Romans with mustaches), then the short-haired (semi-civilized) Gauls, then the long-haired (barbarian) Gauls. Although I have spent a lot of my life in Australia as a long-haired Gaul, I have to question my right to claim that now. If I am honest with myself I need to acknowledge that I can't remember the last time I ate turtle outside of a funeral feast, as a way of living rather than a remembrance of people and times lost. My feet, hands, and belly have become soft, and I use the term *neoliberalism* far more often than I use the word *miintin* (turtle). I may think to myself, "Oh, it's the season to dig turtle eggs and yams now, and the wild pigs feasting on those things will have really good fat. I should go for sugar bag (wild honey) now too." But I'm standing on a train commut-ing to work in Melbourne because I don't have the patience and discipline to languish in a work-for-the-dole program in a remote community, waiting to chase pigs on the weekend. I have to admit I'm something of a short-haired Gaul.

But think about it: which Gaul would a Roman talk to when seeking Indigenous Knowledge solutions to the crises of civilization? Of course, the Romans did no such thing, which may help explain why their system collapsed after only a thousand years or so, but if they had, which Gauls would have offered the solutions they needed? The long-haired Gauls might have shown them how to manage the grasslands and horse herds in perpetuity, but without knowledge of the demands of empire—the grain dole or land entitlements for veterans—their advice would have been interesting but inapplicable. The toga-wearing Gauls would be the right people to ask about the true nature of outsourced tax collection in the provinces (although you might have had to torture them a bit first), but they benefited so much from kickbacks and rewards for suppressing their own culture that they would have contributed little by way of Indigenous Knowledge solutions.

The short-haired Gauls, on the other hand, carried enough fragmentary Indigenous Knowledge and struggled enough within the harsh realities of transitional Romanization to be able to offer some hybridized insight—some innovative sustainability tips to the doomed empire occupying their lands and hearts and minds.

Of course, simplistic categories that rank occupied peoples by degree of domestication do not reflect the complex realities of contemporary Indigenous communities, identities, and knowledge. They certainly do not work in Australia.

Our complex history as Australian First Peoples does not align with most criteria demanded for authentication and recognition by colonists. The Indigenous "self" that has been designed by outsiders to render programs of self-determination safe does not reflect our reality. Even our organization into discrete "nations" (to negotiate the legal structures that facilitate mineral extraction) does not reflect the complexity of our identities and knowledge. We all once had multiple languages and affiliations, meeting regularly with different groups for trade, joining in marriage and customary adoption across those groups, including some groups from Asia and New Guinea. I know that, for many people, elements of those laws and customs are still in place, and I am one of those people.

But I also know that the horrific process of European occupation resulted in the removal of most of us from our communities of origin, many to reserves and institutions far from home as part of forcible assimilation programs. Biological genocide was attempted through large-scale efforts to "breed out" dark skin, with the infamous Stolen Generations representing only one part of this policy. For many women, marrying or submitting to settler males so that their children might pass for white was the only way to survive this apocalypse, while waiting for a safer time to return home.

So the recently imposed "authenticity" requirement of declaring an uninterrupted cultural tradition back to the dawn of time is a difficult concession for most of us to make, when the reality is that we are affiliated with multiple groups and also have disrupted affiliations. For many people, these traumatic relations are unsafe to talk about, while for others there are reclaimed connections that are too precarious to declare.

How might we identify and utilize the various sets of Indigenous Knowledge scattered throughout this kaleidoscope of identities? Not by simplistic categorization, that's for sure. Through the lens of simplicity, historical contexts of interrelatedness and upheaval are sidelined, and the authenticity of Indigenous Knowledge and identity is determined by an illusion of parochial isolation, another fragment of primitive exotica to examine, tag, and display. There are zealous gatekeepers on both sides, policing, suppressing. Most of the knowledge that gets through this process is reduced to basic content, artifacts, resources, and data, divided into foreign categories, to be stored and plundered as needed. Our knowledge is only valued if it is fossilized, while our evolving customs and thought patterns are viewed with distaste and skepticism.

I can't participate in this one-sided dialogue between the occupiers and the occupied. For a start I'm not *manth thaayan* (someone who can speak for cultural knowledge). I'm a younger sibling, so that role is not available to me in our custom. I can speak *from* the knowledge but not *for* it or about all the details. However, I can talk about the processes and patterns I know from my cultural practice, developed within my affiliations with my home community and other Aboriginal communities across the Australian continent, including Nyoongar, Mardi, Nungar, and Koori peoples.

Our knowledge endures because everybody carries a part of it, no matter how fragmentary. If you want to see the pattern of creation, you talk to everybody and listen carefully. Authentic knowledge processes are easy to verify if you are familiar with that pattern—each part reflects the design of the whole system. If the pattern is present, the knowledge is true, whether the speaker is wearing a grass skirt or a business suit or a school uniform.

So I turn the lens around.

I'm not reporting on Indigenous Knowledge systems for a global audience's perspective. I'm examining global systems from an Indigenous Knowledge perspective. The symbols that follow help to express this core concept as a hand gesture:

A reader might understand the physical gesture as a living text by mimicking this image, with the left hand sideways with closed fingers, representing a page or screen or print-based knowledge in general, and the right hand with fingers spread out like a rock art stencil, representing the oral cultures and knowledge of First Peoples. The gesture involves placing the splayed hand in front of the eyes, providing the lens through which to view the closed hand.

This is the basic perspective we will use in this book. To avoid losing it in a void of print, I have built every chapter on oral culture exchanges: a series of yarns with diverse people who all make me feel uncomfortable. I yarn with those people because they extend my thinking more than those who simply know what I know. Some of them I'll name, but many would rather not be captured in print and pinned down to a particular moment of thought, preferring to dwell privately in the generative cultural practice of yarning. Yarns are like conversations but take a traditional form we have always used to create and transmit knowledge.

For each chapter, I carved the logic sequences and ideas arising from these yarns into traditional objects before I translated them into print. I did this to prevent my oral culture perspective from becoming fragmented and warped as I wrote.

For example, for this introduction I spent a couple of seasons making bark shields with my brother-in-law Hayden Kelleher and a Worimi artist called Adam Ridgeway. Adam and I yarned through all my concerns about writing this book and how I would need to protect myself. We cut bark from red gum trees in the right season when the sap was running and the wombats were moving about and the lyrebirds were mating. We shaped the bark on the fire and attached handles to make thick shields. Adam used some of these for an art exhibition where he made creation patterns from light reflected off broken mirror shards stuck onto the shields. He also drew on his iPad the hand symbols I just showed you. So the ideas of this written introduction are in the shields. I simply hold those objects and translate into print parts of the knowledge I see there.

This is my method, and I call it *umpan* because that is our word for cutting, carving, and making—it is also the word now used for writing. My method for writing incorporates images and story attached to place and relationships, expressed first through cultural and social activity. My table of contents is visual, and it looks like this:

Each chapter will include some "sand talk," invoking an Aboriginal custom of drawing images on the ground to convey knowledge. I can't share a lot of the symbolic knowledge because it is either restricted (by age, birth order, gender, mastery levels) or appropriate only for a specific place or group—for example, while Brolga lore might be relevant for me as an Apalech clan member or for others with the same totem, it is not generalizable for all readers. So the knowledge I will share in the sand-talk section in each chapter will be entry-level. It may reference some stories but won't tell them completely. However, I will tell parts of a big story, a metastory that connects and extends all over Australia through massive songlines in the earth and sky, a Star Dreaming that Juma Fejo from the Larrakia People wants to share with all peoples. It goes everywhere that turtles go—and there are turtles all over the world, even in desert country, so it connects everybody.

Juma and I—us-two—have been working with this knowledge and connecting up those stories across the continent since 2012, the year a lot of people thought the world would end because of some weird interpretation of the Mayan calendar. I will include some parts of Juma's Star Dreaming in each chapter to help with deeper understandings of the concepts. There are six of these images, three at each end of the turtle shell, which will be accompanied by a yarn. The seven other images are of my own design, created over a number of years before commencing my doctoral studies because I was worried about my academic knowledge overtaking my cultural knowledge. I needed to produce something in my own way first that was a greater work than a thesis. I have shared these ideas with people in many different places to help them come into Aboriginal ways of thinking and knowing, as a framework for the understandings needed in the co-creation of sustainable systems.

I have been to many conferences and talks about Indigenous Knowledge and sustainability, and I have read numerous papers on the topic. Most carry the same simplistic message: First Peoples have been here for x thousand years, they know how to live in balance with this place, and we should learn from them to find solutions to sustainability issues today. (I often wonder whom "we" refers to in this statement.) They then offer some isolated examples of sustainable practices before colonization, and that's it. The audience is left wondering, "Yes, but how? What insight does this offer today, for the problems we are experiencing now?"

These questions remain unanswered because Indigenous participants are usually offering formulaic self-narratives and cultural artifacts as a window for outsiders to see into a carefully curated version of their past, and the view is one-way.

We're not sharing what we see when we look back through that window. There's a welcome ceremony at the start and a dance at the end, and everybody goes home happy but none the wiser.

We rarely see global sustainability issues addressed using Indigenous perspectives and thought processes. We don't see econometrics models being designed using Indigenous pattern thinking. Instead we are shown a dot painting and implored to make sure we include Indigenous employment in our plans to double a city's population "sustainably" within a couple of decades. Any discussion of Indigenous Knowledge systems is always a polite acknowledgment of connection to the land rather than true engagement. It is always about the *what*, and never about the *how*.

I want to reverse that phenomenon. I want to use an Indigenous pattern-thinking process to critique contemporary systems and to impart an impression of the pattern of creation itself. I want to avoid the ubiquitous Indigenous literary genre of self-narrative and autobiography, though I will include some anecdotes and yarns when examples are needed. What I say will still be subjective and fragmentary, of course, and five minutes after it is written it will already be out-of-date—a problem common to all printed texts. The real knowledge will keep moving in lands and peoples, and I'll move on with it. You'll move on too. Already, you might take away the hand gesture shown earlier, add your own shades of meaning, share it, and grow something from that pattern that could never be imagined on a page. I need to pass these concepts on so I can leave them behind and grow into the next stage of knowledge. Failing to pass it all on means

I'm carrying it around like a stone and stifling my growth, as well as the regeneration of the systems I live in. I'm getting tired of being a middle-aged boy in my culture.

This book is just a translation of a fragment of a shadow, frozen in time. I make no claims to absolute truth or authority. I change gears from academic to campfire voice from moment to moment. Things may seem unstructured; I allow the logic to follow the complex patterns I'm trying to describe, which don't reflect the usual cause-and-effect relations of print-based thought. Words may be capitalized that are not usually capitalized, and this changes in different contexts when they have different shades of meaning. One of the exciting things about the English language is that it is a trade creole, so it changes shape wherever it goes. I will be honoring this quality by taking her for a spin to see how she goes around some tight bends.

This will be a challenge because English inevitably places settler worldviews at the center of every concept, obscuring true understanding. For example, explaining Aboriginal notions of time is an exercise in futility as you can only describe it as "nonlinear" in English, which immediately slams a big line right across your synapses. You don't register the "non"— only the "linear": that is the way you process that word, the shape it takes in your mind. Worst of all, it's only describing the concept by saying what it is not, rather than what it is. We don't have a word for nonlinear in our languages because nobody would consider traveling, thinking, or talking in a straight line in the first place. The winding path is just how a path is, and therefore it needs no name.

One man tried going in a straight line many thousands of years ago and was called *wamba* (crazy) and punished by being thrown up into the sky. This is a very old story, one of many stories that tell us how we must travel and think in free-ranging patterns, warning us against charging ahead in crazy ways. So it will be stories, imagery, and yarns that will make the English work in this book, with meaning being made in the meandering paths between the words, not in the isolated words themselves.

There are many English words to describe our First Peoples, and since none of them are entirely appropriate or accurate, I randomly cycle through most of them, each of which is somebody's preferred term and somebody else's offensive label. Before European occupation we just called ourselves "People" in our own languages, but because I'm not speaking for any single language group, I use many of the inadequate English terms when I need to refer to us collectively. I use many other terms that I don't particularly like, such as "Dreaming" (which is a mistranslation and misinterpretation), because a lot of the old people I respect, and who have passed knowledge on to me, use these words. It's not my place to disrespect them by rejecting their vocabulary choices. I know and they know what they mean, so we might as well just use those labels. In any case, it is almost impossible to speak in English without them, unless you want to say, "suprarational interdimensional ontology endogenous to custodial ritual complexes" every five minutes. So "Dreaming" it is.

I discuss some beginner's knowledge about Aboriginal cosmology, and then look for patterns and implications for sustainability, in a free-range ramble that should never be taken at face value. I write to provoke thought rather than represent fact, in a kind of dialogical and reflective process with the reader. For this I often use the dual first person. It is a common pronoun in Indigenous languages but not present in English; that's why I translate it as "us-two," my fingers typing those letters while my mouth is saying *ngal*.

Solutions to complex problems take many dissimilar minds and points of view to design, so we have to do that together, linking up with as many other us-twos as we can to form networks of dynamic interaction. I'm not offering expert answers, only different questions and ways of looking at things. While I'm good at stimulating connective thinking, I'm certainly no authority on any of the ideas in this book, and my point of view is marginal, even in my own community. But there is fertile ground at the margins.

The hope is this: that from this liminal point of view us-two might be able to see some things that have been missed, glimpse an aspect of the pattern of creation, and run a few thought experiments to see where that pattern takes us. It worked for Einstein, who seldom set foot in a lab but simply said, "If this, then this, then this," creating simulations in a Dreaming space to produce proofs and solutions of startling complexity and accuracy. In this space, even what he thought of as his greatest mistake later turned out to be his greatest discovery. It can't be that hard. If we get stuck, we'll ask the echidnas for help.

We need to begin with the first questions that always form a barrier to approaching this knowledge. Who are the real Indigenous People? Who among them carries the real Indigenous Knowledge, and what aspects of that knowledge are relevant in grappling with the design of sustainable systems today?

Albino Boy

Us-two go walking on songlines with Clancy McKellar, Wangkumarra Song Man. These are ancient paths of Dreaming etched into the landscape in song and story and mapped into our minds and bodies and relationships with everything around us: knowledge stored in every waterway and every rock. We walk the corner-country where Queensland, South Australia, and New South Wales meet. He identifies my ancestral lines and shows me where those stories connect with his.

Neither of us is particularly dark-skinned, and it is perhaps for this reason that he highlights the albino characters in his lore. A white owl woman with fair skin and blond hair who becomes a *Gubbiwarlga*, or clever woman, and is eventually turned into a quartz stone. An albino boy who is ostracized by vexatious elements in his community and banished. When we get to the rock site built by the albino boy, it takes my breath away. He didn't mope around during his banishment—he worked hard, and he didn't skip leg day.

Massive carved and polished rocks are placed throughout the site, lifted by the boy to balance upon standing stones, stacked in piles, or lined up to form processions. There are more than us-two can count over this massive site, which includes a sundial calendar marking the seasons and movements of celestial bodies. I can't understand why I hadn't heard of this place before—why it isn't as famous as Stonehenge. I place my hand on one of the rocks, and there is a deep *duum* that rises through it from the ground, reverberating up through my shoulder and down into my gut. I think I just got the answer to my question.

This isn't an archaeological site, to be excavated and observed. It is still inhabited. The boy is still here, and he probably doesn't want uninvited visitors. It's no monument. The place is alive. Every rock is animate and sentient—but in our worldview this is true of all rocks. Far away there is a secret cave with a miniature replica of the site built on the cave floor. People with the knowledge of how to work with the stones there are said to be able to travel between those sites in the blink of an eye. And these places are connected to stone arrangements all over the continent.

Later, during equinox, I stand at Wurdi Youang in Victoria: a C-shaped stone arrangement that marks the movement of the sun throughout the year. I watch from the sighting stone farther down the hill as the sun sets behind a marker stone at the top of the arrangement, as the moon rises directly behind me, and as Venus, Jupiter, Saturn, and Mars align along that same path. The moment is not just about celestial bodies forming an orderly queue—it is also about a thousand different stories that converge and the pattern they create in a dialogue between earth and sky and me. The way each person knows those stories is subjective—how they are known in that time and place by that person is a unique viewpoint that is sacred, a communication between earth camp and sky camp, between people and a sentient cosmos. Us-two are both there, but we're seeing different stories.

Birds flying overhead are part of that creation song in that moment. A satellite. A plane. Two clouds in the north spiraling strangely, like snakes. We call that a "something," a sign or a message from the Ancestors. I think of Two Snakes story and where I first learned about it, traveling from Gundabooka in northwest New South Wales to the coast. Above me I see Mars and Venus and know them as the eyes of the creator, who in many southern regions sees through the eyes of the eagle during the day, and through those planets at night.

Periodically, a ceremony is held near the border of New South Wales and Queensland, in which Murris bring red opal from Quilpie and blue opal from Lightning Ridge, one from north and one from south, to unite Mars and Venus as the creator's eyes. I think of this, and of the site farther south near Walgett where the eagle eyes are two deep holes in the rock. I think of my woman's totemic relationship to the eagle and how she embodies that connection. I keep expanding out through this web of connections between terrestrial communities and country in the sky. There are living rocks up there as there are down here, and the dark spaces between the stars are not a vacuum, but solid lands that have mass and sentience, reflecting places and times on earth. I can see the pattern—right up until the point I try to write it down, when it disappears like smoke.

There is not much of use here, on the surface of it. These things allow us to assert, "See, we've been astronomers for thousands of years, so our knowledge matters. You wrecked all that, you bunch of bastards." Beyond that, what knowledge can we share to shed light on sustainability and other complex issues? Juma Fejo tells me everything in creation has Dreaming, even windshield wipers and cell phones, so why must our knowledge of creation be frozen in time as an artifact?

Stones in the earth and the sky, all these stories and their connections can tell us more than the mere fact that they have existed for a certain number of millennia. They can tell us about how to deal with the complexities and frailties of human societies, how to limit destructive excesses in these systems, and most important how to deal with idiots. To find this knowledge we need to get practical. We might try some sand talk to get started. Let's look at one of Oldman Juma's symbols.

The two symbols within the hexagon represent different things, more than their recent mathematical meanings. Separately, they are signs of marsupials (<) and birds (>) as different totemic categories of meat, based on the direction their legs bend at the knee. Together (<>) they represent the only two placental mammals native to this continent, humans and dingoes. They form a shape that shows the rules of marriage in a kinship system; on a different angle they can form a Men's Business symbol. They also show a point of impact, a creation event associated with the Orion constellation (always a hunter or warrior everywhere in the world), a big bang caused by Echidna fighting with Turtle. The trauma of this event caused the sky camp and earth camp to separate, and the universe to begin deep cycles of expansion and contraction, like breathing, in a pattern shaping everything.

The big bang pattern, that initial point of impact, is not just something that occurs at the massive scale of the universe but is repeated infinitely in all its lands and parts. Many creation stories refer to this point of impact, often represented by a stone at the center of the place and story. Uluru is the stone at the center of this continent's story, a pattern repeated in the interconnected and diverse stories of many smaller regions, reflected in our own bodies at the navel and then down into smaller and smaller parts at the quantum level of our cosmology. In this way of knowing, there is no difference between you, a stone, a tree, or a traffic light. All contain knowledge, story, pattern. To sit with this story, to discern the pattern, we need to begin by examining rocks.

It would be unhelpful to say, "Granite is an igneous crystalline mix of quartz, mica, and feldspar." It would also be unhelpful to waft around in a tie-dyed shirt hugging the rocks and asking them to divulge their secrets by communing with us through our navel piercings. You have to show patience and respect, come in from the side, sit awhile, and wait to be invited in. So we might do some more sand talk first, before we get to the business of rocks and who is allowed to know about them and how that knowledge might help us survive today.

I spent a lot of time drawing that bird/marsupial symbol over and over, yarning with people about it, and finally making a stone ax to store my understandings. It took me a year. The reason it took so long was that I kept coming back to those two kinds of legs, which formed an image of Emu and Kangaroo in my mind, over and over—as a Dreaming image, but also as Australia's coat of arms. Settlers must have recognized its importance to adopt it as a symbol of their colony. I have unresolved issues with Emu, with its role in creation and the behavioral patterns that keep spinning out from this, making problems for human society and, by extension, all of creation.

Emu's problem can be seen in the mathematical greater-than/less-than interpretation of the symbol. Emu is a troublemaker who brings into being the most destructive idea in existence: I am greater than you; you are less than me. This is the source of all human misery. Aboriginal society was designed over thousands of years to deal with this problem. Some people are just idiots—and everybody has a bit of idiot in them from time to time, coming from some deep place inside that whispers, "You are special. You are greater than other people and things. You are more important than everything and everyone. All things and all people exist to serve you." This behavior needs massive checks and balances to contain the damage it can do.

There are a lot of stories that explain how all this began, and as a Brolga boy (traditional enemy of Emu), I know them all. My favorite one comes from Nyoongar Elder Noel Nannup in Perth, who tells the Dreaming story of a meeting in which all the species sat down for a yarn to decide which one would be the custodial species for all of creation. Emu made a hell of a mess, running around showing off his speed and claiming his superiority, demanding to be boss and shouting over everyone. You can see the dark shape of Emu in the Milky Way. Kangaroo (his head the Southern Cross) is holding him down, Echidna is grasping him from behind, and the great Serpent is coiled around his legs. Containing the excesses of malignant narcissists is a team effort.

The combination of social fragmentation and lightning-fast communication today, however, means we have to deal with these crazy people alone, as individuals butting heads with narcissists in a lawless void, and they are thriving unchecked in this environment. Engaging with them alone is futile—never wrestle a pig, as the old saying goes; you both end up covered in shit, and the pig likes it. The fundamental rules of human interaction do not apply to them, although they weaponize those rules against everyone else.

The basic protocols of Aboriginal society, like most societies, include respecting and hearing all points of view in a yarn. Narcissists demand this right, then refuse to allow other points of view on the grounds that any other opinion somehow infringes their freedom of speech or is offensive. They destroy the basic social contract of reciprocity (which allows people to build a reputation of generosity based on sharing to ensure ongoing connectedness and support), shattering this framework of harmony with a few words of nasty gossip. They apply double standards and break down systems of give-and-take until every member of a social group becomes isolated, lost in a Darwinian struggle for power and dwindling resources that destroys everything. Then they move on to another place, another group. Feel free to extrapolate this pattern globally and historically.

We have stories for this behavior, memorial stones scattered along songlines throughout the landscape, victims and transgressors transformed into rock following epic struggles, standing for all time as cautionary tales. Clancy McKellar took me to a site where three brothers who had kidnapped women were punished and turned to stone. All over that place in Tibooburra the red rocks are people turned to stone for breaking the Law or messing around too much with weather-modification rituals. There is Law and knowledge of Law in stones. All Law-breaking comes from that first evil thought, that original sin of placing yourself above the land or above other people.

In our traditional systems of Law we remember, however, that everyone is an idiot from time to time. Punishment is harsh and swift, but afterward there is no criminal record, no grudge against the transgressor. Perpetrators are only criminals until they are punished, and then they may be respected again and begin afresh to make a positive contribution to the group. In this way, people will not lie and shift blame or avoid punishment by twisting rules to escape accountability. They can look forward to a clean slate and therefore be willing and equal participants in their own punishment and transformation, which is a learning process more than anything else.

This is perhaps something of value to be taken from our stone stories to make justice systems more effective and sustainable today. Those old criminals in stone all over the country are not despised figures but respected entities who received their punishment and are now revered in their roles of keeping the Law. If we respect them and hear their stories, they can tell us how to live together better.

But I don't know very much about rocks. I feel more at home on open savanna and dry sclerophyll bushlands, and my Story Place has only one stone, which moves around of its own accord and so is in a different position every time you go there. It arrived from Asia, carried by a cyclone, and never quite settled down to live slowly like other rocks. So I need to yarn with somebody who really understands the way stone works. As usual, I seek the most insightful knowledge in the most marginalized point of view. I talk to a young Tasmanian Aboriginal boy called Max.

Max has silvery white hair and alabaster skin. He looks and talks like he'd be more at home riding a dragon than a stock horse. He's a proper nerd, memorizing hundreds of digits of pi for no particular reason, thinking his martial arts skills are much better than they really are, and carrying around an encyclopedic knowledge of elves and hobbits and superheroes. He can also write songs in his ancestral language that make me cry.

We've spent a lot of time sparring in a traditional style that was once done with stone knives. The rules of engagement are that you can only cut your opponent on the arms, shoulders, or back (extremely difficult to do) and—here's the kicker—at the end of the fight the winner must get cut up the same as the loser, so that nobody can walk away with a grudge. It's hard enough to cut somebody on the back with a stone knife when they're trying to do the same to you, but it's even harder when you know that every time you cut them you're really just cutting yourself. In our yarns following these sessions we decided this kind of combat forces you to see your enemy's point of view, and by the end of it you can no longer be opponents because you're connected by mutual respect and understanding. More lessons from stone—but how to apply these today? Sounds like a good opportunity for a thought experiment.

I guess if you wanted to take a contemporary economy that is dependent on perpetual war and try to make it sustainable, you could start by applying similar rules of engagement. But in the stone-knife model, enemies are a nonrenewable resource, and eventually you would run out of them. It would not be sustainable at all for the war machine if everybody ended up respecting all points of view. Perhaps the transferrable wisdom here is simply that most young men need something a little meatier than mindfulness workshops to curtail the terrifying narcissism that overtakes them from the moment their balls drop. Maybe then they won't grow up to be the men who start wars in the first place.

This brings us back to that foundational flaw, that Luciferian lie: I am greater than you; you are less than me. Because his appearance does not match some people's idea of his cultural identity, Max faces abusive encounters grounded in that foundational flaw daily. His identity is constantly questioned by both Aboriginal and non-Aboriginal people who place themselves in a greater-than position and get a little thrill out of pronouncing judgment on his existence. Max reflects on these encounters, deciding that these people lack their own authentic identities and therefore can only find comfort in assaulting his.

Max may not know everything about his lineage or his culture, both of which were catastrophically disrupted by large-scale genocide, but he knows who he is, and the fragments of cultural knowledge he carries have integrity and value. He applies the pattern in those fragments to every aspect of his life.

"I don't know what I'd be if I didn't have my identity," he says, "because I haven't really known a life without it. I can't discern parts that are Indigenous and parts that are not because all of my actions are Indigenous—the way I move through the world, my social interactions, my way of thinking about anything. It bleeds through you, no matter what."

When Max recites a hundred digits of pi, he is not stepping outside his identity; he is singing a pattern of creation from north to south. He does not need to have an Elder's level of knowledge to do this. He needs only to perceive the pattern in what he does know. Keepers of knowledge see him behaving in this way and know he is ready to be responsible for additional knowledge, so they pass on story to him. This is how Indigenous Knowledge works.

I'm not saying Max is perfect. He is also prone to moments of the greater-than/less-than deception. He drives me crazy with it sometimes. Once I yelled at him so hard I lost my voice for two days, but then my reaction to his behavior was just as bad, so I had to atone for that. I've taught Max a lot of things, but he teaches me things too, like how to bounce back from a mistake. He also teaches me about rocks, because Tasmanian people have a particular connection to rocks.

"Stones to me are the objects that parallel all life, more so than trees or mortal things because stones are almost immortal," he says. "They know things learned over deep time. Stone represents earth, tools, and spirit; it conveys meaning through its use and through its resilience to the elements. At the same time it ages, cracking and eroding as time wears it down, but it is still there, filled with energy and spirit."

We yarn about the sentience of stones and the ancient Greek mistake of identifying "dead matter" as opposed to living matter, limiting for centuries to come the potential of Western thought when attempting to define things like consciousness and self-organizing systems such as galaxies. Western thinkers viewed space as lifeless and empty between stars; our own stories represented those dark areas as living country, based on observed effects of attraction from those places on celestial bodies. Theories of dead matter and empty space meant that Western science came late to discoveries of what they now call "dark matter," finding that those areas of "dead and empty" space actually contain most of the matter in the universe.

This brings us back to Uncle Noel Nannup's creation story of when Emu went nuts. In that story, the precreation reality was that space was solid: it sat heavily upon the ground, crushing everything that attempted to come into being. Earth and sky had to be separated, the Ancestors lifting up the heavens physically. Sky country is seen in our stories as tangible, having mass, in a way that reveals an understanding of dark matter. All that celestial territory is in constant communication with us, exerting forces upon us and even exchanging matter in the form of rocks crashing through our atmosphere. Our stories show our ancient understanding of the way asteroids form craters, a realization that only entered scientific knowledge a few short decades ago.

Max and I yarn about how our knowledge of these things cannot have always been unique to our culture, as the ancient names for constellations are often the same as ours throughout the world—the Seven Sisters, the Two Brothers, the Eagle, the Hunter. These are global stories and systems of knowledge that must have once been common to all people. We think something terrible must have happened in the north to make people forget, causing science to have to start all over again from scratch rather than building on what went before. What could this cataclysm have been? I imagine the Black Death couldn't have helped, but I suspect it began earlier than that. I think the Emu deception got out of hand somewhere and spread, causing more and more people to think themselves greater than the land, greater than others, greater than the women who hold our lives in their hands and bellies. Whatever it is, this cataclysm is growing, and I wonder how we can stand against it.

Max responds:

Stone teaches us that we should be strong no matter what tries to crack us or wear us down, keeping an unbreakable core through your culture and your beliefs. The majority of this earth is rock, and while water and plants make up its surface, the body of the earth, the part that keeps it all together, is rock. You can have life and creation, but it will all crumble without a solid base. Same with society, companies, relationships, identities, knowledge—almost anything both tangible and intangible. Like those forests and trees sitting as a skin over the rocks of the earth: without that strength inside, without that stone, it would crumble.

Thinking about the shape of the world Max describes and the thin skin around it, I reflect on the physics of our creation stories and the way rocks wear away over time into balls. I perceive a pattern in the universe whereby the most efficient shape for holding matter together is a sphere. I might say to the growing number of flat-earth theorists out there, "Blow me a flat bubble and I'll consider your theory." But that would be placing myself in a greater-than position, so I need to check myself and pay attention to them, remembering that there is always value in marginal viewpoints.

So I listen to them online and realize that the sphere is not the final shape of this creation process. Our own galaxy began as a sphere and flattened into a disk, and the earth is gradually flattening itself too, as it spins like a lump of clay on a wheel. It's only flattened by just over twelve miles at the poles so far, but it's getting there. It's a good thing I didn't dismiss the flat-earthers out of hand; otherwise I might never have understood that properly.

But what use could come from that kind of thinking? Well, a thought experiment might yield a few applications. Packaging, for example, might make more efficient use of space and resources if we considered that you can get a hell of a lot more into a small sphere than a big box. But then what would stop those spheres from rolling off the shelves? The flat-earthers resolve this—just squash the spheres down a bit. Thank you, flat-earthers. That innovation could save a bit of landfill, buy us a little time.

Max thinks it will take a bigger shift in thinking to stave off planetary destruction, that we need to learn more about respect from the stones. I agree. The understanding that we are no greater or lesser than a rock would certainly change things if a critical mass of people all came to it at once. Anyone who thinks they're better than a rock should be turned into one—then they would find out they're not that special, and they could finally be happy. Max suggests that in recent decades people have been becoming aware of rock spirit, reminding me of what has been going on at Uluru.

There is a shed there full of rocks. For a long time, tourists took stones away from that sacred site as souvenirs, and then a few decades ago something strange began to happen. The tourists started mailing the rocks back with panicked reports of weird happenings, disturbed sleep, bad luck, ghostly visitations, and terrible accidents. Somehow they knew it was because of the rocks, and they sent them back with desperate apologies. So many were returned that they had to build a big storage shed to house them.

In our Law we know that rocks are sentient and contain spirit. You can't just pick one up and carry it home, as you will disturb its spirit and it will disturb you in turn. If you sit at any campfire for a yarn with Aboriginal people anywhere on this continent, you will be sure to hear a cautionary tale about a relative who was silly enough to pick up a rock and take it home, who then got sick or was haunted or killed or went crazy. A lot of rocks are benevolent and enjoy being used and traded, but you have to follow the guidance of the old people to know which ones you can use. Rocks are to be respected.

Perhaps further work needs to be done on what constitutes consciousness and what constitutes life. If the definitions of these things could include rocks as sentient beings, it would go a long way toward stemming the Emu-like behaviors that are running rampant across the earth and cyberspace right now. Either that, or we could start mailing those Uluru rocks out to all the narcissists to give them a lesson in respect for others.

Hopefully I have now given you some ideas on what Indigenous Knowledge is, which Indigenous people have it, and what it might be used for. In case you missed them, the answers are everything, all of us, and anything. But who is Indigenous? For the purposes of the thought experiments on sustainability in this book, an Indigenous person is a member of a community retaining memories of life lived sustainably on a land base, as part of that land base. Indigenous Knowledge is any application of those memories as living knowledge to improve present and future circumstances.

First Law

"Why can't you see the flowers?" the young girl said to me. I was teaching her this image in the sand, teaching her about time and deep time, and she humbled me, made me see a different way, a beautiful way that was right now, where now-time was all-time and filled with joy. Oldman Juma made me see it a different way too, put my cheek to the ground and see the "ant's view," the topography of Country in the image on the sand, all the valleys and ridges. Time and place the same. But those three circles—creation time to Ancestors' time to contemporary time—the pattern repeats in the micro as well, with three generations, life stages, and even three parts within a day, a task, a moment.

It comes up off the ground at you, this image, in 3-D, and it is a moving, swirling energy system. And at the outside, a new circle that is really just the one in the middle again, because it all goes back, feeds back into that system in time and place, endless cycles of increase and renewal. We don't die; we go back to Country, then come around again, third time around. Creation time isn't a "long, long ago" event, because creation is still unfolding now and will continue to unfold if we know how to know it.

It all comes out from that central point of impact, that big bang expanding and contracting, breathing out and in, no start or finish but a constant state where past, present, and future are all one thing, one time, one place. Every breath ever taken is still in the air to breathe. I breathe the breaths of the Ancestors, and everybody else's too. Always was, always is, always will be. And there are flowers here, and they make me smile.

Sometimes it is hard to write in English when you've been talking to your great-grandmother on the phone, but she is also your niece, and in her language there are no separate words for time and space. In her kinship system every three generations there is a reset in which your grandparents' parents are classified as your children, an eternal cycle of renewal. In her traditional language she asks you something that translates directly into English as "What place?" but actually means "What time?" and you shift yourself into that paradigm reluctantly because you know it will be hard as hell to shift back out of it again when you go back to work. Kinship moves in cycles, the land moves in seasonal cycles, the sky moves in stellar cycles, and time is so bound up in those things that it is not even a separate concept from space. We experience time in a very different way from people immersed in flat schedules and story-less surfaces. In our spheres of existence, time does not go in a straight line, and it is as tangible as the ground we stand on.

> In the sand-talk image on page 37 there is the point of impact at the center, the creation event that set time/space in motion. It shows three great ages of deep time in rippling rings, all coming into being at the same moment. The symbol creates a three-dimensional apple form in the shape of the universe if you can change your "look" into "gaze" and take it in. The sphere is moving, because the three ages are constantly moving from the outside to the center and then back, expanding and contracting, rolling in on themselves over and over and reproducing in an infinite, stable pattern. This is a sustainable system.

Nothing is created or destroyed; it just moves and changes, and this is the First Law. Creation is in a constant state of motion, and we must move with it as the custodial species or we will damage the system and doom ourselves. Nothing can be held, accumulated, stored. Every unit requires velocity and exchange in a stable system, or it will stagnate—this applies to economic and social systems as well as natural ones. They all follow the same laws.

There are three arcs (or petals, as I now think of them) around the center circle that show the way our social system is mapped onto the creation pattern, with three generations of strong women around every child—sisters/cousins, mothers/aunties, and grannies. The granny's mother goes back to the center and becomes the child, and all of them cycle through those roles forever, the spirit of the child being born back through the land. Each one also occupies all of the roles simultaneously—so the sister is also somebody else's aunty, and grandmother to her niece's daughter.

In this way the system itself is different according to the relational context of the person who is seeing it at any given moment. If you are the child at the center, you see one set of relations, but if you are understanding your own child's point of view, you see another set. The child's aunty is also somebody's child, at the center of her own system. Every time you meet someone and establish your relationship to that person, you are bringing together multiple universes. There is no way to be an outside observer of this system— you have to place yourself in it in order to see it in three dimensions, and you must move around and connect within it in order to see multiple other dimensions. From the outside it is just a flat image.

In contemporary science and research, investigators have to make claims to objectivity, an impossible and god-like (greater-than) position that floats in empty space and observes the field while not being part of it. It is an illusion of omniscience that has hit some barriers in quantum physics. No matter how hard you may try to separate yourself from reality, there are always observer effects as the reality shifts in relation to your viewpoint. Scientists call this the uncertainty principle.

I'm a novice with this physics, but my understanding is that when you are looking for the location of a subatomic speck it becomes a particle, and when you are measuring its movement it becomes a wave, and so its physical reality changes in relation to how you're looking at it. The popular response to this has been, "If I can change reality with my mind, I want the universe to give me a Lamborghini."

Not how it works, I am told when I yarn with Percy Paul, a First Nations man and theoretical physicist at Canada's Perimeter Institute. He seems to feel that the complicated equations of uncertainty have little bearing on his lived reality as an Indigenous person. Listening to his explanations of his own way of being and understanding the universe, I try to put myself in his point of view: I start to imagine that an electron forms a field of probabilities for its potential location at any given moment and can't be pinned down to a single location in linear time, so it forms like a beach where each grain of sand is a possibility of its presence. To me this suggests that tangible reality only exists in defiance of linear time.

I feel silly suggesting this to him, so my ego prevents us from yarning about it. Egos always get in the way of a good yarn. Instead Percy and I talk about the first and second laws of thermodynamics, and he shares some amazing ideas with me, but then our thought-paths diverge and the yarn is done, and we never quite get back to it. In the Indigenous world, you can't push people to share knowledge—you just accept what they think you're ready for. Usually knowledge-keepers will withdraw if they sense narcissism in you, and I know I've approached this yarn in the wrong state of mind. Still, I gratefully pick up the seeds he leaves me.

My yarns with Percy lead me to revisit Schrödinger's cat, which appears to be the best way to help the uninitiated understand the uncertainty principle. In this famous thought experiment, you imagine putting a cat in a box with some poison. You don't know if it has died yet because you can't see it, so in that moment the cat is simultaneously both alive and dead. The act of observing the cat breathing would make it alive, and the act of seeing its sightless eyes frozen in a death mask of agony and panic would make it dead. Jesus.

From an Aboriginal cosmological point of view, the uncertainty problem is resolved when you admit you are part of the field and accept your subjectivity. If you want to know what's in the box so bad, drink the poison yourself and climb in. After my yarns with Percy, I begin to see the uncertainty principle not as a law but as an expression of frustration about the impossibility of achieving godlike scientific objectivity.

Scientists currently have to remove all traces of themselves from experiments, otherwise their data is considered to be contaminated. Contaminated with what? With the filthy reality of belongingness? The toxic realization that if we can't stand outside of a field we can't own it? I don't see science embracing Indigenous methods of inquiry anytime soon, as Indigenous Knowledge is not wanted at the level of *how*, only at the level of *what*, a resource to be plundered rather than a source of knowledge processes. *Show me where some plants are so I can synthesize a compound and make drugs out of it!*

I can't see any sustainability solutions in thought experiments about felines in boxes, but I do see possibilities in the pattern created by those three generations of women I mentioned earlier. They show me that you have to move and adapt within a system that is in a constant state of movement and adaptation. By extension, this is also how us-two might influence the system in sustainable ways—any attempt to control the system from a fixed viewpoint outside is a misaligned intervention that will fail. So how might a society respond to global financial instability, for example? Us-two could run a thought experiment from that great-granny/niece story, in a simulation of the universe experienced by those three generations of strong women who form the sustaining social fabric of an extended family.

Through their eyes, we might perceive, throughout all of the massive economic upheavals of recent decades, a few small islands of sustainability and stable growth linked to one thing. No, it's not gold. It's extended family. The third world's remittance economy—which is basically the money sent home by people who have migrated to first-world nations, and which rivals international aid in scope—did not crash during the last financial crisis. In fact, in many instances it grew, baffling economists. These masses of desperately poor people working away from home continue to send billions of dollars back to their extended families and communities, and this economic process turns out to be resistant to contraction.

Meanwhile, in France and Germany, some aspects of the economy relating to extended families also remain stable, just as they have managed to do through two world wars and the collapse of European empires. This is due to communal property ownership laws that do not exist in the globalizing system of the Anglosphere. French and German extended families are able to hold capital collectively and run midsize family businesses without any individual nominally owning and controlling it all. These form permanent intergenerational estates. They work together with diverse portfolios and pool multiple incomes to ensure their eggs are never all in one basket. They provide an internal welfare safety net for each other as protection against random incidences of austerity and upheaval.

Perhaps Australia and other countries could begin depression-proofing its economy by introducing extended family property ownership laws and incentives—as a bonus, this might also ease the welfare burden and decrease unemployment. If we're uncertain what such a model might look like, we might employ consultants from Australia's Asian community, which seems to be already operating an informal economy based on extended family models. Or we could just ask one of my grannies.

This chapter is supposed to be about physics, I know, but you can no more separate fields from one another than you can separate yourself from the field. No matter what field you are in, everything is nature and therefore follows the same natural laws, the same physics. The universe from your point of view may be different from everyone else's, but all follow the same laws.

First Peoples and Second Peoples, however, seem to have a fundamental disagreement on the nature of reality and the basic laws of existence. First Peoples' Law says that nothing is created or destroyed because of the infinite and regenerative connections between systems. Therefore time is nonlinear and regenerates creation in endless cycles. Second Peoples' law says that systems must be isolated and exist in a vacuum of individual creation, beginning in complexity but simplifying and breaking down until they meet their end. Therefore, time is linear, because all things must have a beginning, middle, and end.

Aristotle invented that idea. For him, the end (telos) is the principle of every change. It is a strange kind of curse that demands an illusion of infinite growth based on inevitable decrease and annihilation. Second Peoples and their captives are required to believe wholeheartedly in this paradox, which is only possible through the I-am-greater-than fallacy, in order to ignore the First Law and experience time in a straight line.

Us-two can't blame Aristotle for this, however. The idea was already present in the form of the foundational civilizing mythology of the ouroboros. This was a metaphor representing infinity—a snake in a circle, eating its own tail. However, it contained the same curse, the same contradiction: How can this serpent be a symbol of infinity if it will eventually eat itself?

I composed this chapter by carving a *boondi*, or wooden club, from a mulga tree. In my clan we would call it a *yuk puuyngk*, or Law stick. I thought it was an appropriate medium for exploring the different laws of time and space for First and Second Peoples. I studied the laws of thermodynamics and yarned with Elders and Percy Paul about these things (along with a bunch of old dead white guys), then stored that knowledge on my inner maps of the Great Dividing Range, which is the body of the Rainbow Serpent. It divides nothing, by the way, but connects systems along a massive songline. Parallel is the Great Barrier Reef, another serpent in carpet snake form, which is a barrier to nothing, by the way, but another infinitely connective story. I have traveled that songline with Elders and knowledge-keepers, from Caboolture to Hinchinbrook Island. Indigenous Knowledge is kept in such songlines, so that's where I stored that data, which I also carved into the club as a mnemonic to help me remember it.

Around the head of the club I etched an image of the ouroboros to represent Second Peoples' law and the second law of thermodynamics. But the three-dimensional nature of the club gives this two-dimensional image an additional layer of meaning, a truth that is revealed when you roll the stick across clay. An image appears of an endless procession of snakes, head to tail, representing the First Peoples' Law and the first law of thermodynamics.

In the first law of thermodynamics, energy is neither created nor destroyed—it only changes and moves between systems. In the second law of thermodynamics, entropy or decay increases in a complex system as it inevitably breaks down, giving rise to what physicists call "the arrow of time"— but only in a *closed system*. Perhaps the desire to create closed systems and keep time going in a straight line is the reason for Second Peoples' obsession with creating fences and walls, borders, great divides, and great barriers. In reality we do not inhabit closed systems, so why choose the second law of thermodynamics to create your model of time?

When us-two see that arc in the sky, that Rainbow Serpent, we are only seeing one part of it, and it is subjective: just for us. If we move, the rainbow also moves, only appearing in relation to our standpoint. If you go to the next hill, you will see it in a different position from where I am seeing it. The moon sisters were trapped by a similar phenomenon, chasing the reflected moon on the surface of the night sea, thinking it was a fish they could spear. But like the rainbow, that image moves in relation to where you are sitting, so they could never catch it. Now you can see their shadows in the moon where they remain trapped to this day, a warning to all about chasing the illusion of fixed viewpoints.

The Serpent loves the water because that is what allows us to see him, and he communicates with each of us this way, but he is not just an entity of water. He is an entity of light. The part we are seeing there in the wet sky, or in the fine spray coming off the front of a speeding dinghy, is just a line across the edge of a sphere. The line moves across multiple spheres that are infinitely overlapping, spiraling inward and outward, extending everywhere that light can go (or has gone or will go), and the Rainbow Serpent moves through this photo-fabric of creation. He goes under the ground too, because light has been there in the past and he is not limited by linear time.

Ah, but is he a wave or a particle? I guess that depends on how you're looking at him, but we would see him as a wave, a snake, because he is constantly in motion across systems that are constantly in motion and interwoven throughout everything that is, was, and will be. There are infinite variations of him in all shapes and sizes throughout the world—

wyrm, dragon, uraeus, and many different names in different regions, taking the shape of the spirit of those places. He was always there and always will be, unless people keep trying to make him eat his own tail.

I can't see him properly, though, because I'm color-blind. (Optometrists tell me this is because I'm not a "full-blood"— apparently, the blacker you are, the less likely you are to be color-blind.) I can only see the Serpent as a vague, thin streak in the sky. My color blindness, however, makes me look elsewhere for him, finding knowledge in unexpected places. My impairment also allows me to see the camouflaged snake in the grass in front of us as we walk and yarn, so it's a useful disability. Every viewpoint is useful, and it takes a wide diversity of views for any group to navigate this universe, let alone to act as custodians for it. I stand in this gully and see the Rainbow Serpent in one place; you stand on the hill and see him in another, and he gives us different messages that we are supposed to share with one another.

My subjective view of the Rainbow Serpent helps me perceive problems with the timelines we are all forced to inhabit today (although it also makes me miss appointments and write in logic sequences that can be difficult to follow). The arrow of time proposed by physicists works in lab experiments and is a real, observable phenomenon in closed systems. It is a true law. It's just the wrong law to apply to beings living in open, interconnected systems. It's a bit like touting the theory that an economy is thriving when the stock markets are doing well—the actual inhabitants of the economy say, sure, stock prices are spiking, but we're still hungry!

The selective application of different laws and theories

is the reason for the crisis of civilization that will be experienced on this planet until we reach Aristotle's telos, the inevitable end. It is a metaphor based on deception, and in an Aboriginal worldview this is how curses work. You take part of a system (e.g., a person's hair) and observe the pattern of the whole system (e.g., his or her body, mind, and spirit) in that fragment, then you sing a false pattern (e.g., you are dying) into the whole from the part, powering it with another system in constant motion (e.g., running water). The curse is a deception made real—either an outright lie or a true law or pattern applied where it doesn't belong. It is like a computer virus, a sneaky line of code that ends up crashing the whole system. The application of the second law of thermodynamics to open, interconnected systems does the same thing. It is a curse.

To understand the crisis of civilization in this way, we first need to define what civilization is from the standpoint of First Peoples' Law. Many would say it is a culture that produces knowledge, technology, law, and arts, but this could be said of any community in the world. Indigenous people will often say, "We had these things too, so we had a civilization!" But a civilization is something else. A village or pastoral community or mobile community seasonally managing ancestral estates is not a civilization, because civilizations build cities. Wakanda in Marvel's Black Panther comics is an African civilization, because it makes cities. In the real world, the ancient peoples of Zimbabwe who once made cities of stone lived within a civilization, until it inevitably collapsed. This was not an Indigenous culture just because its inhabitants had dark skin. Civilizations are cultures that create cities, communities that consume everything around them and then themselves. They can never be Indigenous until they abandon their city-building culture, a lesson the Elders of Zimbabwe have handed down from bitter experience through deep time.

A city is a community on the arrow of time, an upward-trending arrow demanding perpetual growth. Growth is the engine of the city—if the increase stops, the city falls. Because of this, the local resources are used up quickly, and the lands around the city die. The biota is stripped, then the topsoil goes, then the water. It is no accident that the ruins of the world's oldest civilizations are mostly in deserts now. It wasn't desert before that. A city tells itself it is a closed system that must decay in order for time to run straight, while simultaneously demanding eternal growth. This means it must outsource its decay for as long as possible.

For this reason, a city is dependent on the importation of resources from interconnected systems beyond its borders. The city places itself at the center of these systems and strips them to feed its growth, disrupting cycles of time and land and weather and water and ecological exchange between the systems. The exchange is now going only one way. Matter and energy are still neither created nor destroyed in this reaction; they are directed into static heaps rather than cycled back through and between systems.

The exponential destruction caused by cities feeds the exponential growth of infrastructure and population. For this they misapply laws like supply and demand: in order for economic growth to occur, there must be more demand than supply. Roughly translated, that means there must be more people needing basic goods and services than there are goods and services to meet their needs. Put another way, there must be a lot of people missing out on what they need to survive in order for the economy to grow or in order for anything to have value. As the growth continues exponentially, so do the masses of people missing out. There is no equilibrium to be found here.

You need to stave off disruptions from those desperate masses with bread and circuses, football and Facebook. You need to fragment them so they are not supporting one another in communities or extended families; otherwise your demand base decreases. Above all, you need them to breed like rabbits, so you make sure their only asset is the potential energy of their children.

I don't think most people have the same definition of sustainability that I do. I hear them talking about sustainable exponential growth while ignoring the fact that most of the world's topsoil is now at the bottom of the sea. It is difficult to talk to people about the impossible physics of civilization, especially if you are Aboriginal: you perform and display the paint and feathers, the pretty bits of your culture, and talk about your unique connection to the land while people look through glass boxes at you, but you are not supposed to look back or describe what you see.

But that First Law is still there. We need to be brave enough to apply it to our reality of infinitely interconnected, self-organizing, self-renewing systems. We are the custodians of this reality, and the arrow of time is not an appropriate model for a custodial species to operate from. If I think about all those grannies and nieces and sisters now, I wonder if I haven't gone the wrong way charging down all these wormholes of physics and poking every negative particle. Those women just quietly get on with things and keep creation systems in motion through kinship, and they don't worry about much, except what kind of mess I might make next. Maybe they've got this. In a lifeworld where your great-grandchildren become your parents, you have a vested interest in making sure you're co-creating a stable system for them to operate in and also ensuring a bit of intergenerational equity. So, in quiet moments, I just like to sit in the bush within the comfortable embrace of those women's spirit of creation. I can still hear the bulldozers coming, and I can no longer hear the frogs. But I can see the flowers.

Forever Limited

Us-two will now consider for a moment the end of the world and the limits of "forever," via some nonlinear thoughts about Santa Claus, flags, racial illusions, maps, and fishing boomerangs. As usual, we begin with a yarn.

I sit with two Sami women, Indigenous people from near the North Pole who have a sacred totemic connection with the reindeer herds they still follow today across arctic landscapes they have inhabited for thousands of years. They remain connected to these lands, which are for the moment claimed by temporary modern states such as Finland, Sweden, Norway, and Russia. They are tipping my world upside down.

They are wearing modern clothing and speaking in accents that seem Nordic to my untutored ear. They have yellow hair and rosy cheeks. And yet, despite the evidence of my senses, there is no part of me that is able to recognize them as "white." I have exactly the same feeling sitting with them as I do when I sit with aunties in my own community. There is the same pattern of logic, of being, of connectedness. Their way of relating to me is a mirror of my own way of relating to other Indigenous people. But they are senior knowledge-keepers in their community, and I am just a young fella who is still trying to understand his place in the world. I may have darker skin than these ladies, but they are ten times blacker than me.

When the yarn is over, I find that "whiteness" is no longer a useful term in my vocabulary. In my community we use the words "black" and "white" every day as a convenient shorthand to describe relationships between the occupiers and the occupied, but those terms are horribly inadequate for describing our reality today, particularly in multicultural and international contexts.

In a world where black African colonists are annexing the traditional lands of fair-skinned Nemadi hunters, where Celts struggle against English domination while Basques in Spain and Koryaks in Russia fight to retain their ancient lands and languages, where diasporic peoples of various skin tones have been making babies together for generations in every country, black and white is a limiting paradigm for understanding the Indigenous experience. The Indigenous experience itself is hard to define as a distinct reality when the non-Indigenous communities of the world have only (relatively) recently been displaced from their own ancestral territories of origin, moved into large cities and towns to provide the labor that progress demands.

It is difficult to name the ripples and patterns of global power systems when we are limited by nineteenth-century language around race and colonialism. For Australians, the hazy old binaries of race have become profoundly unsettling and difficult to pin down on a color-coded continuum of victims and oppressors. "People of color" in their struggle for economic equality join the rush to exploit Aboriginal land and resources, and they are welcome at the boardroom table as long as they embrace settler values and identities. An Indian company undertakes a project to devastate Aboriginal lands and waters in Queensland with coal mining, and farmers formerly opposed to Indigenous land rights activism now stand beside the Aboriginal community to protest the development. African American visitors are offended when they drop in on Indigenous centers in our universities and hear us using the term "black" to describe ourselves, when so many of us can no longer scrape together enough melanin to scare off a taxi.

The Sami women I'm yarning with are unable to use the archaic shorthand of "white" to name the people who are setting up rocket-launching facilities on their lands for space exploration, drilling for oil on their sacred sites, killing off their reindeer and language and old people while poisoning or damming their rivers. How can these perpetrators be named? Are specific cultural groups responsible? Are individuals?

What group or individual could take responsibility for the West appropriating Sami culture in the most barbaric ways at Christmas time? Their shamans were generous enough, early last century, to share with outsiders some spiritual practices involving red and white fly agaric mushrooms in that season, only to have elements of those rituals stolen and co-opted into European culture in sacrilegious ways.

Their traditional pointed hats were ripped off and placed on childlike elves. Santa's camp was moved from Holland to the North Pole, and the shamanic Sami practice of entering dwellings through the chimney (to distribute psychotropic mushrooms) was adapted for his use. He changed the color of his costume to match the iconic red and white of the Sami mushrooms. Then he took off with their reindeer and gave them stupid names, using them for transport as he expanded the global reach of his holiday empire. If the Sami apocalypse had a soundtrack, it would be "Jingle Bells."

Can we blame one aging, overweight "white" male for this cultural theft? Can we blame all old, fat white men? Perhaps we could consider the possibility that the destruction of lands and cultures globally involves a little more than people deciding individually whether they will play the role of hero, villain, or victim. Perhaps even ethnic groups can no longer be categorized accurately this way.

When I finish my yarn with the Sami women, there is a huge gap left in the vocabulary I have been using to describe my world. The words my community was given in the language of the occupiers to describe our racialized reality have blinded us to the true nature of that reality. We name victims and perpetrators by a color code that masks the actual forces and patterns that are vexing us. We see ethnic groups with our left eye and individuals with our right eye, and we are blind to anything else. We can't see the flows of power and control in the world, the systemic suck of resources from south to north. An illusion of time and space, a false global map of places and peoples has slid over our frontal lobes.

As Aboriginal people, we have always had our own ways

of mapping time/place. Perhaps, to fill the gap those Sami women have created in my worldview, I need to revisit our maps and extend the patterns they contain into worlds beyond the local. So us-two might look at Oldman Juma's map in the sand: let's follow those patterns out and see where that leads us in identifying the unsustainable patterns and forces that are threatening all of existence.

This is both a map and a kind of compass, but it is not aligned with magnetic north. In the old way, direction is dynamic and based on seasonal solar movement from your point of view from where and when you are standing, walking, or camping at any given moment.

Time and place are usually the same word in Aboriginal languages—the two are indivisible. At the center of the compass is the point of impact during a creation moment or site; this and the other points represent seven spirit families and their sacred places.

The first man and first woman are at the east and west, with their lines of travel shown, creating different kinship groups and sites as they move through Country. The travel lines on the inside arcs of the symbol form the shape of the first man's canoe. The lines of travel on the outer arcs show how he paddles to turn it, as the sun turns, making direction more dynamic than the modern magnetic version. This is enough to hurt your mind. "What? North is not a fixed magnetic point? East can be north? It changes depending on the seasons?" I recommend being dismissive at this point, unless you want to take a couple of days off work to lie down and reset your brain. That seems to happen to a lot of people in this particular bit of sand talk.

It gets worse when Oldman says that this view of the symbol is a flipped image that you can only see properly if you send your mind under the ground to look at it from beneath.

Together, all the lines in the sand-talk map show the shape of the Australian continent, which is an upside-down version of the Mercator version we all know today. In many non-Western languages, including Maori, Middle Eastern, and Aboriginal languages, north is down, and south is up. Early maps of the world were like this too, before Europeans began their empire-building and inverted the charts to place themselves at the top. They also stretched the top half of the map to make themselves look bigger, so that the equator is actually lower than it should be and tiny Greenland looks like a continent. If you don't believe me, look at any map and see where the equator is. It is not in the center, but below it. In Oldman Juma's map, Tasmania is at the top of the world, and it is a lot bigger.

I decided to yarn with someone from the top of the world, a Tasmanian Indigenous thinker called Lauren (aka Blackie), who lives with her woman in Sydney and is working on her doctorate. She has done some good writing about development and its intersections with Indigenous realities. I carved a fishing boomerang out of ti tree wood and gave it to her a year ago. It has Oldman Juma's maps of time and place on it, along with some images of Ant and Butterfly from a dance we once worked on together. A fishing boomerang is a rare weapon from Western Australia that most people have never heard of.

When I got permission over there to carve them a few years back, I made heaps of them. They don't look like regular boomerangs—they are about the size of an outstretched hand and shaped a little like an angular comma with a sharp point. Oldman Juma calls them blackfella ninja stars. They fly very fast and zip into the water to knock out fish. If it is made right, the angle of the weapon helps you judge where the fish are, rather than where they appear to be as the water bends the light. There are good lessons to be learned from using those boomerangs—lessons about illusions, perspectives, and realities.

Originally they were made from hardwood, but a lot of people started making them out of roofing iron when that material became available last century. I thought this elusive, adaptive instrument was a perfect fit for these slippery schools of ideas about time and direction, which are difficult enough to capture in oral culture modes, let alone print.

I gave this small but deadly boomerang to Blackie to carry in her handbag for protection, as it is statistically not particularly safe to be an Aboriginal woman walking home in the city at night.

I recently asked Blackie what she has been thinking about while carrying that boomerang for a year. She replied:

> When you put your hand over one eye and then the other the whole thing takes off, wings flapping like a butterfly. It changes direction as you look at it with different eyes. The whole symbol lifts off the boomy [boomerang], strong wings flapping and creating updrafts that ripple throughout the rest of the story. I suppose that's part of the transformation. As we encounter these (un)limited growth nations, we really just need to get momentum in those wings and take the fuck off. Lift off from these unsustainable systems. Like you say, these models are always placeless, universal, but they often forget about space too. As resisters we also need to remember about space. I think there's such a resurgence of place-based everything that even blackfellas overlook space, time, motion, dimension.

Blackie talks about decolonizing movements that have been so intent on rejecting Western systems of thought that they have focused too much on ways of knowing rather than ways of being, causing a lot of Indigenous Knowledge to be lost in theory rather than being embedded in daily life. On the other hand, a recent obsession with "ontology" has swung the pendulum back the other way, as people seek authentic but individually unique foundations for the traditional knowledge they report on in various media.

In all of this branding and rebranding of Indigenous Knowledge, things can become lost or contaminated. This is not like replacing wood with roofing iron in the manufacture of fishing boomerangs; that's fine, as it demonstrates continuity and adaptivity in response to change. It's more like somebody making up a Dreaming story about the Japanese visiting Australia thousands of years ago when they hear Oldman Juma joking about blackfella ninja stars. Indigenous Knowledge is constantly under threat from such weird amendments and misinterpretations, from within and without. The physical apocalypse of invasion came with a bang, but our cultural Armageddon is more of a whimper, a gradual contamination and unraveling of communal knowledge by exceptional individuals.

I've seen Dreaming stories invented with no connection to songlines or real places. I've seen complicated Indigenized rituals manufactured from new-age fripperies. One person says, "Today I'm running a *buddubigwan*, which is the traditional Aboriginal word for workshop," and delivers a PowerPoint presentation mixed in with some trust falls, face painting, and warm fuzzy affirmations. Another person translates the *Oxford English Dictionary* into an Aboriginal language, inventing new Indigenous words for things like "tableau," "quixotic," and "xenophobia." This is a work of genius, but the problem is that it is not how cultures adapt and evolve over time. Like all things that last, it must be a group effort aligned with the patterns of creation discerned from living within a specific landscape.

The innovators creating the above examples argue that our culture is constantly changing and adapting, and they are right. But true cultural change doesn't happen unilaterally. Cultural innovations occur in deep relationships between land, spirit, and groups of people. A person "of high degree" in traditional knowledge may find a song in a dream if they are profoundly connected to land, lore, spirit, and community. But that song must then be taken up by the people and modified gradually through many iterations before it becomes part of the culture. Besides, that song can only be found through a ritual process developed over millennia by that community. The song itself is not as important as the communal knowledge process that produces it.

Most lasting cultural innovations occur through the demotic—the practices and forms that evolve through the daily lives and interactions of people and place in an organic sequence of adaptation. When these processes are unimpeded by the arbitrary controls and designs of elevated individuals, they emerge in ways that mirror the patterns of creation.

I have seen authentically demotic innovation at work in my culture and have observed it carefully over time. I know what it looks like when it is true. I was involved in the emergence of a new funerary ritual over the first decade of this millennium, a headstone-opening ceremony that takes place a year or so after a deceased person is buried. It began in a community that had only wooden crosses in the cemetery and was led by a woman who wanted a stone marker erected for her dead son.

The creation of the ceremony to "open" that first headstone was a communal process shaped by multiple Elders and family members, incorporating older elements of the traditional mourning process that had fallen into disuse. When it was repeated and modified by many families and different communities, it became an authentic innovation embedded in living culture. It even incorporated family savings plans and budgeting to save money for the headstone and the community feast following the ceremony. This complex, beautiful, and healing ritual could not have been designed arbitrarily by any individual or even a "working group."

Sustainable systems cannot be manufactured by individuals or appointed committees, particularly during times of intense transition and upheaval. For those seeking sustainability practices from Indigenous cultures, it is important to focus on both ancient and contemporary knowledge of a demotic origin, rather than individual inventions or amendments. That is not to say that all demotic innovations are benevolent. But if you listen to many voices and stories and discern a deep and complex pattern emerging, you can usually determine what is real and what has been airbrushed for questionable agendas or corrupted by flash mobs of narcissists.

So I am careful about the Indigenous Knowledge I choose to present in this book, because I know how ideas can get tangled up and twisted in the marketplace of this civilization, embraced and repackaged and marketed in forms that are often the opposite of the original concept or intent. I may present elements of Indigenous Knowledge to reveal a true understanding of "forever" in this writing, as a call to align with creation as an infinitely complex, self-sustaining system. I may illuminate ways to follow the patterns of creation in the innovation of truly sustainable solutions; but perhaps the worst possible outcome of this work would be civilization embracing these ideas.

Let's say I get paid to consult with mining companies as they lift a raft of phrases and buzzwords from my work to insert in their annual reports and reconciliation action plans. Maybe my consultancy grows into a business proudly called Forever Consulting. Maybe someone makes me an offer I can't refuse and I sell it, then they turn it into a company called Forever Ltd. that goes global. Maybe my writing then becomes a vehicle for Indigenizing and rebranding the face of extractive technologies, which are becoming increasingly unpopular as *Avatar* depression continues to spread among the restless masses.

Avatar depression is a real condition. It arose from James Cameron's 2009 movie about blue alien indigenes being invaded by a mining company, and psychiatrists have been struggling to treat it ever since. Apparently, the vision of land-based living and culture presented in that film struck a deep chord in millions of people who realized what they had given up, that their civilized lives were bleak and gray and mind-numbing. They had to be medicated back into accepting the illusion of progress again, which is why we had to wait so long for a sequel.

The most remarkable thing about Western civilization is its ability to absorb any object or idea, alter it, sanitize it, rebrand it, and market it. Even ideas that are a threat can be co-opted and put to work. The Romans did it with Christianity—an ideology of the poor and enslaved that threatened the foundations of empire. When torture and murder became ineffective as deterrents, they simply embraced the idea and made it the state religion, rewriting the holy texts to suit their needs and rebranding it as a new system of control. In the same way that plants can be tweaked at the genetic level to become the intellectual property of one company and then replace all similar crops in a region, ideas can be reengineered to serve the interests of the powerful. It's not a conspiracy; it's just power doing what power does.

The people may rise up against tyrants in the name of liberty, shattering the halls and towers of the powerful, but then the ruling systems will simply embrace the idea of freedom, tweak it a little, and continue with business as usual. Liberty becomes the right of land-owning white males to vote, then changes form again to include males of every class, then again to include females, and so forth. It constantly shape-shifts, eventually enshrining the freedom of corporations to make messes they cannot be held accountable for, to bribe governments to change laws allowing them to damage people and land at will, no matter who the people vote into government.

In this way, liberalism has been frequently rebranded to vanquish competing ideologies. The success of liberalism lies in its ability to wear whatever shape a population projects through disruption or dissent. Thus at the beginning of this millennium it remains the only show in town. It is an illusion that currently dominates the globe.

This illusion has a pattern. Everybody follows the pattern, even if they openly oppose the tenets of liberalism or the system of nationhood in general. The most roguish of nations still must maintain their status as a nation, and to do so they must follow the blueprint. No matter where you go in the world, you will recognize elements from this template—if you chance upon a place that doesn't have those elements, you'll find that the people there have lost or are in the process of losing their right to exist.

Everywhere you go, there will be the same institutions, anthems, and flags. There will be recognizable schools, banks, hospitals, government bodies, and courts (no matter how poor or rudimentary), and there will be a dreary national anthem. None of these things existed in a universal form a couple of centuries ago, but everyone has them now. The other element you must possess is a flag. It must be rectangular and utilize three colors somehow representing a unifying ideology and national identity.

In Aboriginal Australia, we are terrifyingly close to joining this recent madness as "First Nations." We have hospital-like and school-like community-led institutions, and we have a flag. In our defense, we need to comply with these things to ensure our right to exist. To date we have resisted settling on a national anthem, thank goodness. The flag is interesting, though. It is understood by most of us as a symbol of defiance rather than compliance, as demonstrated by its proliferation in the cheeky graffiti of Aboriginal children. You don't often see other kids scrawling the Aussie flag on a wall instead of a dick and balls. This is because the Aboriginal flag represents a social system in direct opposition to the global order that requires the existence of flags in the first place.

The Aboriginal flag represents more than just black for the people, red for the land, and yellow for the Dreaming. It shows how the relationship between people and land is balanced perfectly by the Laws, stories, and values of the Dreaming at the center of two equal halves, so that the needs of the people are always in a sustainable balance with the needs of the land. I have done a lot of thought experiments using the flag as a kind of symbolic graph, changing the shapes and sizes of the different sections of the flag to represent different economic and political systems.

In most contemporary systems, the red and the black are alarmingly out of balance, and the yellow at the center seldom takes the shape of a circle. The Dreaming of our current growth-based economic system takes the form of a pyramid. Progressive approaches may try to flip that pyramid upside down or cut off its top, but there is still only a thin strip of red land left at the bottom of the flag no matter where you sit on the political spectrum—every civilized system demands growth based on destruction of land. The political spectrum itself is an illusion, suggesting that the only possible forms of social organization are liberalism, fascism, or socialism.

This limiting range of governance paradigms denies the existence of a myriad of forms of human society developed over eons of existence. These new paradigms ignore the fact that it really is not possible to maintain massive nations and cities in any sustainable form. Civilizations over the last few millennia have done unspeakable damage to the systems necessary for existence, but this is nothing compared to what has happened in the last century or so with the emergence of a global system of great industrial nations.

I won't insult you with endless statistics about extinction rates and topsoil loss and climate change and toxicity levels in every breath of air and every drop of water on the planet, statistics that will be out-of-date a few months after I write them as these disasters continue to multiply. Suffice to say that from my birth up until this point, well over half of the vertebrate species on the planet have become extinct. The 20 percent of the planet that has been left as habitat for the remaining species has been reduced to a series of isolated systems that are unable to exchange uncontaminated matter and energy to maintain complexity. No single cause such as global warming can account for this—greenhouse gas emission is only one of a multitude of factors contributing to this elimination of the biosphere.

When it comes to global environmental catastrophe, the jury is done deliberating and the case is closed. It requires no more debate. This apocalypse is real. On the upside, apocalypses have proven to be survivable in the past, although on the downside it usually means that your culture and society will never be the same again. Oldman Juma says all this has happened over and over again and will continue to happen as the universe breathes in and out. You can live with it, but you need to adapt and change every aspect of your society and culture when transitioning from one era to the next.

All over Australia we have stories of past Armageddons, warning against the behaviors that make these difficult to survive and offering a blueprint for transitional ways of being, so that our custodial species can continue to keep creation in motion. Butterfly goes off on her own to opal Country to chase a shiny new brightness, only to become trapped in ice. Millennia later, the ice melts and her colors run down into the opal, while saltwater people all around the continent keep stories to preserve the maps and memory of lands drowned after the big ice melt. The stories are passed down, and people partner with whales, dolphins, and others to continue caring for the Country beneath the sea. This is important, as the oceans will fall again as they have before, and we will want to return to that Country.

Inland on freshwater Country, the Baakindji people faced extinction when they experimented with nation-building long ago, which worked well enough until the land and sky moved and they were no longer able to move with it. Their story, as I was told, recalled a time when all the tribes and clans of the region gathered and stayed in one place in permanent settlement. There were abundant resources to support this lifestyle, and the people assimilated into one uniform language and culture, forgetting their previous diversity.

A massive meteor crashed nearby and killed most of the people, scorching goannas (large monitor lizards) with different marks to make diverse varieties as a reminder to the survivors of the right way to live. Move with the land. Maintain diverse languages, cultures, and systems that reflect the ecosystems of the shifting landscapes you inhabit over time. That is the blueprint, and we are not the only people who know it—you might recall a similar biblical story in Genesis about the Tower of Babel.

There is an undeniable pattern in the sum total of all these old stories from around the world, indicating that sedentary lifestyles and cultures that do not move with the land or mimic land-based networks in their social systems do not transition well through apocalyptic moments. There are recent stories too, from both Indigenous and non-Indigenous peoples around the world, which also fit this pattern.

While I was tracking this pattern, I yarned with Dr. Larry Gross, chair of Native American studies at the University of

Redlands in California. An Anishinabe man of the Minnesota Chippewa, Gross has published and spoken extensively on a theory he invented, called postapocalyptic stress syndrome (PASS), which occurs "when a culture experiences such a massive shock that it never fully recovers."

Gross has identified this syndrome in both Indigenous and settler communities. Recent examples include manufacturing workers in the American Midwest, whose way of life came to an end with the collapse of their industry. Historically, he found the same indicators of PASS in population data from Europe after the Black Death epidemics in the fourteenth century.

"The Europe that came out of the Black Death was not the same as the Europe that went in," Gross told me. He drew parallels between this event and the Indigenous experience of colonization. "Both resulted in an intergenerational pandemic of posttraumatic stress disorder, suicide, and widespread substance abuse."

Gross sees potential apocalyptic crises today as opportunities for reconciliation, drawing from his study of historical Indigenous and European holocausts: "Since the Europeans went through an apocalypse and suffered the exact same symptoms as Indigenous people, this indicates that we are not as different as we might like to think." He said that it takes at least a century after the crisis has passed for a culture to recover from postapocalyptic stress syndrome, but that the emergent culture can never return to what it was before. It always transitions into something new, adapting to the new social and ecological environment.

Considering that the catastrophes we are experiencing

may take decades or even centuries to play out, then another century for us to recover after that, it may be advisable for us to get ahead of the game and begin creating cultures and societies of transition, to lessen the impacts of this calamity on our communities and potentially avoid postapocalyptic stress altogether. We need to start working with the land, rather than against it. Our communities need to share knowledge with one another while maintaining their own unique systems grounded in the diverse landscapes they care for.

Any real move toward sustainability will require us to cease limiting our understanding with simplistic language around group and individual identities, villain and victim branding, so that we can see what our actual diversity looks like and what it can do for us. We might then begin to notice the patterns and forces that are threatening the survival of all living things and start to change the way we do business. Rather than fighting brand wars to make this doomed globalizing system feel more fair and inclusive, we might instead develop some new systems of transition.

Old Nyoongars and Yorgas in Perth tell stories about a group of three totemic entities that work together in miraculous ways. Certain butterflies always lay their eggs on a particular bush above the nest of a particular species of ant. The ants collect the eggs and take them down into the nest. When the larvae hatch, the ants carry them up to eat the leaves of the bush at night and then carry them back down again. When they grow too heavy to carry, the ants bring the leaves down to them. The larvae grow a jelly on their sides when they eat those particular leaves, and this is the food that the queen ant eats. The larvae then spin cocoons in the nest for the final stage of the process, after which they fly out of the nest as butterflies and begin the cycle all over again.

This intensely interrelated process within a totemic group of three entities—bush, ant, and butterfly—would be impossible for a single human mind to design. How do these symbiotic dances develop, when the cause-and-effect relations are so interdependent and complex that there is no way to reverse engineer the process by which the system came to be? This is precisely the kind of process we need to understand and engage with to create sustainable responses to the catastrophes we are facing.

Lines in the Sand

I'm standing at the edge of a lagoon near the small town of Nyngan in central New South Wales, talking to Aboriginal children about their futures. These bright tomorrows go under the rubric of "learning and earning," a program that has been developed elsewhere using a predictive process I am not familiar with. I'm here to sell them that vision: I'm supposed to be a role model, telling them they can achieve anything if they work hard and show up and smile a lot. So I show up for this excursion to Nyngan. I smile and get to work trying to teach black kids in the bush about their bright economic future while grounding them in the proud traditions of their past.

In the sand I draw lines to represent the scene laid out before us across the lagoon and beyond—three lines representing the river, the old disused railway bridge, and the newer highway bridge. But you can't really understand those unless you understand the different eras and economies that used each of them as part of a supply chain, so more lines are added to represent these elements. You also need to know what happened before that; if you map out all the relationships, you might see a pattern that represents the future, because all time is one time. But when I do that, the future I'm seeing and the future I'm selling are two different things.

There is an old rusted fish trap on the bank, left over from mission times. Beside it I sketch out the shifting economies represented in this vignette—traditional river-based economy, to mission economy, to riverboats transporting cotton and wool, to rail doing the same after the river was destroyed, to the highway with oil tankers and trucks filled with low-grade ore buzzing past. From settlement onward, locals in each era believed the system to be stable and planned their futures around it.

They thought, "We will farm sheep!" and invested all their money and training and time in that, raising their children in it. They thought the incredibly abundant pasture would last forever, not being aware that these fields had been carefully cultivated by the Aboriginal people who were custodians of the river country and that the cessation of this custodianship combined with the introduction of rapacious sheep would destroy this resource completely, along with the topsoil. They didn't count on the industry becoming dominated by a few giants that would squeeze them out, either.

Economic refugees from that catastrophe thought, "We will become shearers!" and invested all their training and time and effort in that career, raising their children to follow the family trade. But the work was variable and not well paid. Some got jobs on the docks or on the steamboats shipping wool along the rivers. Children had aspirations of becoming steamboat captains. We can all be steamboat captains if we work hard enough! People were trained and educated and built lives around this as though it would last forever. But when the scrub was cleared and the last of the topsoil ran into the river, it silted up and the steamboats stopped running.

Railway! Human ingenuity to the rescue! There were no longer as many sheep or as much wool. They planted out cotton and other things in the depleted soil, and they brought fertilizer on the trains to coax crops out of the dead ground. Copper was mined and transported. Children had aspirations of becoming train drivers. We can all be train drivers if we work hard enough! People were trained and educated and built lives around the rail and the cotton fields as though it would all last forever. But those train tracks now are twisted, rusted, abandoned. Years later, most had to admit that farming was not viable on this devastated land base, and they left in droves. Suddenly there were agricultural towns in the region with a majority Aboriginal population again. With the alarming potential of these lands returning to their traditional custodians, policing became the only growth industry in the area.

But the next layer of bridge and highway did well for a while. The last of the copper was doggedly scraped from the earth; oil and gas were found further inland, and people became miners riding a boom. Infrastructure projects for the mines would provide jobs forever! Children had aspirations of leaving school as soon as possible and earning more than the school's principal just by holding a stop sign in a mine. No particular skills or knowledge were needed. They bought ATVs and jet skis and discovered new and exciting uses for the choked and muddy river.

At the end of the mining boom, as construction jobs dried up and unemployment rose and interest rates fell, the next generation of infrastructure—towers with broadband signals—offered hope of new lands, cyberspaces to colonize. Children had aspirations of IT careers and moved away to the city, where they served coffee and rubbed the bloated feet of baby boomers while finding entertainment, if not employment, in their electronic devices. In their digital ghettos there were endless new worlds and resources to discover, and this could last forever, without limits. Or at least until the rare earth metals required to power their devices ran out.

I tell those children who are thinking of dropping out of school and getting technical training or trying for a job in a supermarket that these economies fall apart fairly regularly, and you don't want to be on the bottom of them when they do. I look around at their faces, either blank or scared or angry, and I realize that nobody wants to hear this. I need to show them how to read patterns and see past, present, and future as one time and let them navigate the system themselves.

I take another group of young Aboriginal protégés in Western Australia on regular excursions, walking the countryside with Noel Nannup, who is a local Elder here. He says things like, "It is going to rain in twelve minutes," and the kids time it on their phones and laugh in amazement when his prediction comes true. He predicts events like an annual emergence of flying ants from the ground, then follows seasonal signals, winding through the bush with us to stop under a tree, then snaps his fingers—"Now!"—as the ants explode out of the ground in that instant.

What can I do with that? Say it is amazing and ask him to

tell the kids stories about growing up in the bush? Ask him to make us a damper (bush bread) and show us how to throw a boomerang and tell the kids they can follow their dreams and do anything? Nah. I ask him to explain the patterns of his thinking in making predictions and discuss whether the kids could apply those patterns to contexts beyond the bush. So he shows them his process of pattern thinking and even shares how he uses it to follow stock markets and economic trends. His process is all about seeing the overall shape of the connections between things. Look beyond the things and focus on the connections between them, he says. Then look beyond the connections and see the patterns they make. Find the sites of potential risk and increase, like judging where the ball will go in a football game.

Later I do a workshop with those Western Australian children on the monetary system, reminding them of what the Elder showed us. I get them to make a pattern from this symbol:

The pattern over the next hour spreads intricately over a massive sheet of paper, in between reading, viewing videos, and talking about the structure of the monetary system. At the end of the lesson, they have to find patterns in the random, complex image on the paper and align these with patterns they discern in the monetary system. They are a little alarmed at the sustainability issues that emerge in their analysis.

One student in particular develops a high level of understanding of pattern thinking that he can apply to most problems. In another session, he is present on an excursion to a beach that is eroding into the sea and must be fortified with concrete and sandbags to protect the buildings and property there. The children are asked to design an engineering solution to the problem.

It seems as though this boy is not engaging with the task. He stands under a clump of she-oak trees and stares out at the sea while the others draw and build models of walls and spits and elaborate engines. He appears to be a noncompliant student. Misbehaving. Maybe I should punish him, humiliate him in front of his peers until he complies with the work task. He is not achieving outcomes, not delivering against performance indicators to close the gap. I walk over and ask him what is going on. "Well, it's all fucked," he says. Maybe I should rebuke him for inappropriate language. Instead I ask him what he means.

He talks about what he's learned from Pop Noel about the she-oak trees and underground fresh water flowing beneath them where they grow like that on the coast. He points out those flows into the sea and tracks the subtle movements of the sand out there in the tides and currents, tracing the pathways of constant motion all along the coast, infinite white grains swept up and deposited on new beaches in cycles of cleansing and renewal. He points out a spit in the distance that has been built to block that flow and keep the sand on one beach for its residents, noting that new sand can't be deposited here now because of it. He mentions dozens of other constructions like this along the coast, and the dredging of sand further out to sea to deposit on the beaches to maintain them as real estate and public facilities.

Then he turns around and points at the buildings, observing that they are made mostly out of concrete, which is made mostly out of sand, much of which is dredged from the ocean floor, leaving holes and gouges in the seabed that fill up with sand again. He notes that the sand moves around in its cycles but never makes it back to the beach. Or worse, the seabed slumps into those holes and the beach then collapses further into the sea. "You can build all the levies you like," he says, "but those fuckin' buildings are gunna go back into the sea where they came from."

Well. As I always say, if you want to find the next generation of great thinkers, look in the detention room of any public school.

If we sand-talk through the symbol on page 77, drawing it on the ground and then continuing its pattern, we see that it is focusing on the things people usually ignore. People today will mostly focus on the points of connection, the nodes of interest like stars in the sky. But the real understanding comes in the spaces in between, in the relational forces that connect and move the points. This symbol highlights those interconnections and de-emphasizes the points. If you can see the relational forces connecting and moving the elements of a system, rather than focusing on the elements themselves, you are able to see a pattern outside linear time. If you bring that pattern back into linear time, this can be called a prediction in today's world.

For this yarn I made a boomerang etched all over with this symbol. Trends and surprises emerge within the whole design, and what seems like chaos has patterns and shapes that you can only discern with a holistic view. Contemporary science is beginning to understand this way of knowing through chaos theory, complexity theory, network theory, and fractal geometry. It is becoming clear that complex systems are adaptive, self-organizing, and patterned with a logic that can be discerned and used for trend analysis and predictive technologies. Second-wave automation, artificial intelligence, and blockchain technologies seek to harness this complexity. But it is a complexity that cannot function through external design and control.

You could create a mechanical ballet by programming a group of rolling robots to move in sync. But your program would be a closed system designed from the outside—one robot would get bumped an inch out of place, and the whole thing would fall apart. You can do it cheaper and faster just by letting a dozen of those little vacuum cleaner robots loose together in a space to move randomly. At first they will just clean the floor individually, but over time maybe the rhythms of creation will begin to move them together, as they start to dance in patterns that are more startling than anything a single programmer might design, like schools of fish in the sea.

Consider boids, which are digital objects that move on a screen. When you place a number of boids on a screen with three or four simple rules telling each to match the velocity of the others, move randomly, and avoid collisions, after a while they will begin to move together as a group in complex patterns resembling the flocking of birds. These patterns cannot be programmed but must emerge within the system organically—a process that is called "random" in Western worldviews but is in fact following the patterns of creation.

Similarly, effective artificial intelligence is not created simply through programming—it is more efficient to program a simple AI and let it loose on a complex system of data. As it moves through the data, it becomes a learning entity and programs itself autonomously. A blockchain cannot be designed externally as a closed system, or it will stagnate. It must comprise individual nodes that remain autonomous, operating freely in a self-organizing system of users. The internet also developed in this way. The kinds of digital innovations that are currently disrupting top-down global economic and social structures are built on the reality of complex, self-organizing systems rather than the illusion of centralized control.

This has implications for the management of all systems, particularly social-control systems. Community members, like boids, birds, fish, or nodes, need to operate autonomously under three or four basic rules, self-organizing within groups, spaces, and data sets to form complex learning communities. The patterns and innovations emerging from these ecosystems of practice are startling and transformative and cannot be designed or maintained by a single manager or external authority. They cannot even be imagined outside of a community operating this way.

This is the perspective you need to be a custodian rather than an owner of lands, communities, or knowledge. It demands the relinquishing of artificial power and control, immersion in the astounding patterns of creation that only emerge through the free movement of all agents and elements within a system. This impacts the way we are managed and governed.

Preindustrial cultures have worked within self-organizing systems for thousands of years to predict weather patterns, seasonal activity, and the dynamics of social groups, then manage responses to these complexities in nonintrusive ways that maintain systemic balance. While interventions are possible from *within* these dynamic systems, they cannot be controlled from the outside. Systems are heterarchical—composed of equal parts interacting together. Imposing a hierarchical model of top-down control can only destroy them. Healthy interventions can only be made by free agents within a complex system—agents referred to in chaos theory as "strange attractors." Could you be a strange attractor within your institution? It is a risky endeavor in a culture that attaches negative meanings to words like "chaos" and "anarchy," equating them with disorder and ruin. But chaos in reality has a structure that produces innovation, and "anarchy" simply means "no boss." Could it be possible to have structure without bosses?

In my community there is a phrase that is repeated daily: "Nobody boss for me!" Yet at the same time, each person is bound within complex patterns of relatedness and communal obligation. Indigenous models of governance are based on respect for social, ecological, and knowledge systems and all their components or members. Complex kinship structures reflect the dynamic design of natural systems through totemic relationships with plants and animals. Totems can also include other elements of these systems, like wind, lightning, body parts, and substances. The whole is intelligent, and each part carries the inherent intelligence of the entire system. Knowledge is therefore a living thing that is patterned within every person and being and object and phenomenon within creation.

Respectful observation and interaction within the system, with the parts and the connections between them, is the only way to see the pattern. You cannot know any part, let alone the whole, without respect. You cannot come to knowledge without it. Each part, each person, is dignified as an embodiment of the knowledge. Respect must be facilitated by custodians, but there is no outsider-imposed authority, no "boss," no "dominion over." While senior people ensure that the processes and stages of coming to higher levels of knowledge are maintained with safety and cohesion, there is no centralized control in Aboriginal societies. Western knowledge systems are centralized, and this could be why they have so far been unable to engage in dialogue with Indigenous Knowledge systems in the development of sustainability solutions. It could also be part of the reason I have so much trouble bringing Indigenous Knowledge processes into dialogue with the academy—the other part being that people are too busy filing reports, performance reviews, and funding proposals in these institutions to have any time for complex Indigenous Knowledge. It is far easier to collect, bag, and tag samples from our culture. But nobody says that.

Mostly academics tell me that they are unable to incorporate Indigenous Knowledge into the academy because their students are not smart enough to understand it.

This is frustrating. It is accepted that in the academy we study examples of knowledge mastery and the finest thinking available in our field. There is no reason the same standard should not be applied to Aboriginal Knowledge, which is no more or less accessible than ideal gas equations or genetic sequences. No matter how difficult a topic is, there are always simple points of entry and general principles and processes to be gleaned by novices encountering it for the first time. We have the same thing in Indigenous cultures—stages of knowledge and no progression without mastery and respect. It just requires a bit of discernment, humility, and awareness. It is easy enough to observe, listen, and gradually increase your participation as your mastery increases.

When I first began to read Dante back when I was twenty, I didn't need in-depth knowledge to start engaging with his unfamiliar ideas and cultural perspective. Within an hour of my first encounter with his *Inferno*, I had mapped out all the layers of hell and decided which level each of my enemies was headed for. It made me feel so much better. Thank you, Dante! After that I went overboard and spent a year learning the Florentine dialect so I could read the original text, but I can't say this in-depth study helped me understand European people any better.

To be honest, I was really only studying Dante and his dialect to show off and try to impress college girls. It worked sometimes, but never with the European ones. There's a lesson there. First, check your ego and your motives. Why are you doing this? Second, you don't need to be an expert to understand the knowledge processes of people from other cultures and enter into dialogues with them. More important, making yourself an expert in another culture is not always appreciated by the members of that culture.

Understanding your own culture and the way it interacts with others, particularly the power dynamics of it, is far more appreciated. My reading of Germaine Greer when I was a young lad was a lot more conducive to forming relationships with European females than my reading of Dante was—and that was more about understanding my male privilege and controlling its excesses than being an expert on women's literature or issues.

This kind of cultural humility is a useful exercise in understanding your role as an agent of sustainability in a complex system. It is difficult to relinquish the illusions of power and delusions of exceptionalism that come with privilege. But it is strangely liberating to realize your true status as a single node in a cooperative network. There is honor to be found in this role, and a certain dignified agency. You won't be swallowed up by a hive mind or lose your individuality— you will retain your autonomy while simultaneously being profoundly interdependent and connected. In fact, sustainable systems cannot function without the full autonomy and unique expression of each independent part of the interdependent whole.

Sustainability agents have a few simple operating guidelines, or network protocols, or rules if you like: diversify, connect, interact, and adapt.

Diversity is not about tolerating difference or treating

others equally and without prejudice. The diversification principle compels you to maintain your individual differences, particularly from other agents who are similar to you. This prevents you from clustering into narcissistic flash mobs. You must also seek out and interact with a wide variety of agents who are completely dissimilar to you. Finally, you must interact with other systems beyond your own, keeping your system open and therefore sustainable.

Connectedness balances the excesses of individualism in the diversity principle. The first step in connectedness is forming pairs (like kinship pairs) with multiple other agents who also pair with others. The next step is creating or expanding networks of these connections. The final step is making sure these networks are interacting with the networks of other agents, both within your system and in others.

Interaction is the principle that provides the energy and spirit of communication to power the system. This principle facilitates the flow of living knowledge. For this, you must be transferring knowledge (and energy and resources) with as many other agents as possible, rather than trying to store it individually. If the world ever experiments with an actual free market rather than an oligopoly, this would be the perfect system to facilitate sustainable interactions. Knowledge, value, and energy in truly sustainable networks of interaction are prevented from remaining static and unchanging by the final protocol.

Adaptation is the most important protocol of an agent in a sustainable system. You must allow yourself to be transformed through your interactions with other agents and the knowledge that passes through you from them. This knowledge and energy will flow through the entire system in feed-back loops, and you must be prepared to change so that those feedback loops are not blocked. An agent that is truly adaptive and changing is open to sudden eruptions of transformation, in which the agent may temporarily take on the role of strange attractor and facilitate chain reactions of creative events within the system.

You can see these principles embedded throughout Aboriginal culture. I can see many of them just looking at our first-person pronouns. In English these are "I"/"me" and "we"/"us." In Aboriginal languages there are many more, including pronouns that are translated as "I," "I myself," "we two," "we but not others," and "we altogether." Repeating the plural ones twice can mean "It's up to us," but repeating "I" twice can mean "I go my own way!" There is a balance between self-definition and group identity. These two are not contradictory but entwined, and there are names for all of the roles you occupy as an agent of complexity in Aboriginal society. You perform these roles alone, in pairs, in exclusive groups, and in networked groups. Our languages are expressions of land-based networks and facilitate communication across all of these individual nodes and collectives of nodes within and between systems.

In Aboriginal English there is a very useful term to help us draw lines in the sand between these roles, asserting both boundaries and connections. The word is "lookout" (a term I believe we originally borrowed from Cockneys), which is not a warning but refers to a person's appropriate sphere of influence and accountability. If a person is being pushy and expecting you to get involved with his frantic business in an effort to control you, you might say, "That's not my lookout," or "Nah, that's your lookout." Your lookout encompasses all of your reasonable obligations and activities within your pairs, groups, and wider networks. By reasonable, I mean any tasks that reflect the other protocols of a sustainability agent—leaving you free to be different from others, receive and transfer knowledge, and transform in response to shifting contexts, while acting as a custodian and defender of these things.

You have to keep moving, both as a strange attractor and as a sustainability agent. If you are following the four protocols of working within the pattern, little systems of vibrant complexity will spring up around you, and other strange attractors will draw you into their amazing webs. These will attract other agents, but they will also attract narcissists in droves. You need to insulate yourself against these entropic elements, but you also have a responsibility to help them.

We have looked at Emu story, but it is also worth looking at stories from the northern hemisphere to discern the patterns of narcissism that prevent the functioning of sustainable systems. The word "narcissist" comes from a Greek story about Narcissus, a man who fell in love with his own reflection in the water. A girl called Echo was in love with him, but she was cursed to hang around and only repeat his words forever. This is what you find with these narcissist flash mobs: one loud person will start shouting silly things and attract followers who repeat those things without thought. Not all strange attractors are benevolent.

Narcissism isn't incurable though. Survivors of this plague emerge without any memory of who they really are, needing support to begin again and relearn the nature of their existence, their purpose for being here. They are like children, and leaving them to their own devices at this stage is not advisable. Entire cultures and populations recovering from this plague have been left like orphan children with no memories of who they are, longing for a pattern they know is there but can't see. They grow up eventually, but it takes a long time if they have no assistance. There are so many adolescent cultures in the world right now, reaching for the stars without really knowing what they are.

Adolescent cultures always ask the same three questions. *Why are we here? How should we live? What will happen when we die?* The first one I've covered already with the role of humans as a custodial species. The second one I've covered above, with the four protocols for agents in a complex dynamic system. The third one us-two will look at next.

Of Spirit and Spirits

That word just doesn't look right in print. It looks alien, weirdly Egyptian or something. I have to keep checking the spelling to make sure I've got it right. S-P-I-R-I-T. And it's just one word to describe so many things that I need to write about—*ngeen wiy, maany, oony way, ngangk pi'an*, and so forth—things we have so many words for that have no translation in English. English words like "ghost," "revenant," "haunt," "force," "soul," and "essence" don't quite capture what I am talking about. Us-two will have to find metaphors, stories, and analogies to make this yarn work.

I made a pointed parrying shield out of *thaanchal* wood while thinking about all this, to represent the protection needed while navigating the spaces between the tangible and intangible worlds that custodial species must engage with. And now I need to add some new layers to the story, because of where I am choosing to write it. I did not bring the shield with me, and there are ghosts in this place.

Well, there are ghosts all over the massacre-soaked continent of Australia, but they generally don't do much harm
unless you fall asleep right on top of them or mess with their
places and things. More specifically, there are settler ghosts
here where I am writing, and they're angry as hell.

I won a fellowship to do a writing retreat for a week at
Varuna, the heritage-listed home of one of Australian literature's seminal authors, Eleanor Dark. "Seminal" doesn't
sound like the right word to describe a female's work. Ovarian? Ovarian writer in the Australian canon? No, stuff it,
women have always kept the seeds, chosen the seeds, sown
the seeds. Seminal. Her work is seminal.

My yarns for this chapter are with dead people like her
and with a random group of writers, also staying at Varuna,
who are experiencing some ghostly happenings. Let me
make it clear: I'm not actually communing with the dead
here, beyond yelling at them to go away when they get too
close at night. (Okay, maybe squealing like a tiny child is
more accurate.) Normally I'd just smoke the place out with
smoldering leaves to get rid of the ghosts, but I can't really do
that in this protected heritage site.

So I'm in the Blue Mountains where they build dams and
go-kart tracks over our Aboriginal heritage places, desecrat-
ing a songline for Rainbow-Serpent-Eel, but I'm being careful
to respect the heritage status of this famous desk and study
I'm sitting in. I won't write her name again because I feel her
around here, and I worry if she sees me typing it she'll attach
herself to me and make me sick. I swear she's reading over my
shoulder right now.

She had a hard life, the old *wadjin* (white lady). The other settlers didn't really play fair with her. She was a Marxist at a time when that wasn't very popular, in an era when people didn't tweet nasty things about you but refused to sell you bread and lamb's liver and suet or whatever they used to eat. She had it tough all right—eventually it got so bad that she had to move to another one of her properties in another state (poor thing only had a few properties). She seems nice enough, and I don't think it's her tormenting us.

I'm sleeping in the maid's room (poor thing had only one maid), and all the trouble seems to start there. I don't think the entity is female. It is something in the house that is not happy to be creeping down the stairs (in the middle of the night to gramophone music) and then tiptoeing into the maid's room to find a hairy blackfella lying on the bed.

On the second night here I went for a shower to escape the disturbance, only to be somehow knocked out cold. I woke up on the bathroom floor with blood everywhere. Six stitches and a concussion vigil later, the other authors upstairs, all female, are experiencing things. One keeps getting her hair stroked by some invisible hand, and at those moments the nearest wall becomes spongy and elastic if she tries to touch it. Another spent two nights with something sitting on her chest so that she couldn't move.

They've all stayed here before but never experienced these problems in the past, although they've heard stories. Toilets flushing on their own. A ghostly mustached lip nuzzling the necks of female residents. I think the spike in paranormal activity during this residency is my fault. The ghost expected a maid in the maid's room, but the maid had a thick black beard in this case and made him feel confused about his ecto-sexuality. Now he's all unbusted ghost nuts and acting out.

I'm going to stop typing for a bit and try to find a place to make a sneaky fire and smoke myself. Maybe I'll ask some of the ladies if they want to get smoked too, so we have an easier night tonight. It's hard making an unsanctioned fire for cultural purposes on Country though—you may get arrested or fined for it. Still, even a night in the lockup would be better than another night here without the protection that smoke offers.

○

That's done now, just smoked myself and one of the other writers, the one who has been crushed and held down during the night by that restless spirit. She's still leaving today though—she needs to get some sleep. I feel cleaner, as if all those sticky bits of settler ectoplasm have been dissolved off the edges of my spirit.

Here's how the smoke works. It is made by the leaves: light from sky camp and nutrients from under the ground, connecting the two worlds and moving between them, visible but intangible. You have to feel it going through you, through your body and particularly through the big spirit at the center of your belly. The smoke is liminal—neither earth nor air but part of both—so it moves across the same spaces in between as shadow spirits do, sending them on their way.

The shadow spirit is that part of a person that collects attachments to things, sensations, places, and people. Some First Peoples in New South Wales call it *yaawi*, which early settlers adopted as a name for their bogeyman—yowie. It is all longing and illusion, the part of your spirit that carries the I-am-greater-than delusion. *I am special*, it screams and is drawn to its own name and image. This is why in Aboriginal cultures we often won't say the names of the deceased, or any words that sound like their names, and will cover or hide any photos or images of that person. Our word for this spirit is often the same as the word for image. It thinks it will live forever, that its temporary persona represents full consciousness and being. It is pure narcissism.

All around the world in the original cultures of humanity there are similar rites to assist this spirit to dissipate and fade. There are death wails as part of the grieving process, songs to sing, smoke to spread, and a sequence of mourning that should take place in stages over a year. But most people don't have these things anymore, with new and vague stages of grieving laid out for us that contain unhelpful steps like Denial and Bargaining. There may be memorial shrines or web pages with photos of the deceased and much calling of their names. As a result, these shadow spirits linger years longer than they are supposed to, tormenting our nights with their whispered claims to exceptionalism.

There are at least four parts to your spirit from an Aboriginal point of view, and this shadow is only one of them. Your higher self (maybe what they call the "superego" in psychology) is your big spirit, and it goes back to sky camp when you die. But sky country always reflects earth country, so there is another spirit, your ancestral spirit, that goes back to a place in the land. It is born again eternally from that place. There is at least one other part, your living spirit, which animates your body in life, flowing through you from the land around you like water fills a string bag in a running creek—never the same water in the bag from moment to moment. That water is only as good as what is in the creek. Therefore if the land is sick, your living spirit is sick as well.

Your shadow spirit is that part of you that wants things you don't need and makes you think you're better than other people and above the land, and it takes all the other parts of your spirit to hold it in check. If the rest of your spirit is not clear and in balance, it gets away from you, causing conflict and destruction. You gossip behind people's backs, spread uncertainty, deliver judgments, upset people, take more than you need, and accumulate goods without sharing. It makes you a competitor instead of a human being. But only when it is out of balance. If it is checked by the other parts of you, it becomes a stable ego that drives you to act upon the world in perfect ways.

You don't need to believe in ghosts to balance spirit and live the right way in this world. You can use any metaphor you like—for example, ego, id, superego, and persona. Frontal lobe, monkey brain, neocortex, and lizard brain. Athos, Porthos, Aramis, and d'Artagnan. Harry, Ron, Hermione, and Malfoy. Monkey spirit, Pig spirit, Fish spirit, and Tripitaka. Matthew, Mark, Luke, and John. Whatever stories your cultural experience offers you, you can still perceive spirit through metaphor and bring it into balance to step into your designated role as a custodian of reality. Some new cultures keep asking, "Why are we here?" It's easy. This is why we're here. We look after things on the earth and in the sky and the places in between.

The circle on the left represents the abstract world of mind and spirit, and the circle on the right represents the concrete world of land, relationships, and activity. The lines above and below show the lines of communication between these worlds, which occur through metaphors. These metaphors include images, dance, song, language, culture, objects, ritual, gestures, and more. Even written words are metaphors that help carry communication between the abstract and practical realms (although that communication usually only goes one way and does not complete the loop shown in this image). Metaphors are the language of spirit. They go around, top and bottom, because you need to close the feedback loop—you can't just sit in the abstract space, because you need to take the knowledge back to apply it in the real world, and vice versa. This can be seen in a secular view of reality as a relationship between theory and practice.

The sand-talk symbol shows a basic model of the Turnaround event of creation, the enormous revolving force that produced the separation of earth and sky worlds. "Turnaround" is an Aboriginal English word that was used to describe creation events and times before the term "Dreamtime" was invented by settlers. Creation is not an event in the distant past, but something that is continually unfolding and needs custodians to keep co-creating it by linking the two worlds together via metaphors in cultural practice. Story Places or sacred sites are places of overlap between the two worlds, which is why people need protection when entering these places—calling out for the old people and putting armpit sweat or smoke or water on those entering. Ceremony creates a similarly powerful overlap between the worlds. Ceremonies and interactions with sites on Country in this way keep creation in motion, causing increases in natural and social systems that are necessary for good health.

A smaller but similar Turnaround event happens at the neurological level when an individual learns something new. There is a spark of creation like lightning when true learning takes place, with a genetic reward of chemical pleasure released in the brain. This is the moment that teachers love—described by educators universally as "the light coming on in their eyes." You can see the same light when you gut a fish—for a few minutes there is a shine like rainbows in its intestines, but as the life and spirit leave those organs, the light dies. This living spirit of creation, sparked by opposite fields colliding and separating, is what brings fire and light into the universe. This is the sacred nature of knowledge. A knowledge-keeper must share knowledge because she or he is a custodian of miniature creation events that must continually take place in the minds of people coming into knowledge.

The chemical burst of pleasure we feel when genuine knowledge transmission takes place occurs from the creation of new neural pathways. These are connections between two points that were previously unconnected. Jokes are one of the most pure examples of this neural creation event; most humor is based on two ideas coming together in a new way: puns, rhymes, double meanings, unusual circumstances, accidents, exposed delusions, and contextually inappropriate content are examples of this. The chemical rush we get from sudden neural connections in jokes is so intense and pleasurable that we laugh out loud. This kind of humor and joy in learning is a huge part of Aboriginal and Torres Strait Islander cultures. If people are laughing, they are learning. True learning is a joy because it is an act of creation.

But there are two kinds of joy. One is characterized by lightheartedness, and the other is marked by fierce engagement and deep concentration. Both give pleasure by increasing connectedness and complexity in the neural systems of learners. There needs to be an interaction between abstract (spirit) and concrete (physical) worlds of knowledge for this kind of complexity to develop fully. Without closing the loop between abstract knowledge and reality, and without making connections between different ideas and areas of knowledge, true learning cannot occur.

A focus on linear, abstract, declarative knowledge alone not only fails to create complex connectivity but damages the mind. We are biologically punished for this destructive behavior with a neurochemical rush of lethargy and discomfort that most people call boredom. Extended periods of this affect a person's mental health, resulting in bouts of rage, depression, and worse. In centralized knowledge institutions today, this illness is called misbehavior or misconduct. Without the spark of creation in your neural system, the mind-body system stagnates and falls apart, affecting not only your ability to learn but your health and relationships as well, leading to increasingly destructive behaviors. If you are an Aboriginal person living in Australia, this will almost certainly lead to incarceration and/or a decreased life span.

The creative spark is a process that allows us to solve seemingly impossible problems. It involves representing real-life elements with metaphors, which transform tangible things into spirit, images in an abstract space. That is the action of the line at the top of the symbol. In that Dreaming space, the abstract entities can be manipulated and reorganized to find solutions to real-life problems. This is how the thought experiments in this book have been conducted. For a simple example of this Turnaround process, consider the way four apples and two hungry people can be translated into the abstract symbols of 4/2. The solution (=2) may then be found in the abstract space, transferred back to reality and applied to share the apples equally. If one person has more nutritional requirements than the other, then the abstract of 3:1 might be applied.

Mathematics is widely dreaded by most people the world over because of the recent tradition of confining its operations to the abstract/theoretical world. Without connecting math to real-life contexts, people feel damage being done to their neural systems and naturally resist. There needs to be meaningful schematic links made between the symbols and what they may represent in our lived reality.

I have observed math classes conducted by my colleague Dr. Chris Matthews (an Aboriginal mathematician) in which corroboree dances have been expressed as mathematical equations, and then new equations have been formed and new dances created to express them. What made these rituals effective was not simply the cultural content of the dances—it was the Dreaming action of translating a real-life event into metaphor, then manipulating the metaphor to gain understanding, followed by innovation transferred back to the real world. Traditional culture is important, but it is not just a performance or display—the Dreaming process is the key. The same process applied to a spreadsheet or a birthday party would be just as effective. The key to Aboriginal Knowledge, as always, lies in the processes rather than just the content. Token inclusion of cultural clippings serves only to further diminish and exclude the cultural identities of First Peoples. This damage to culture and identity can also damage the mind.

The mind extends to the nontangible world, the left-hand circle of the sand-talk diagram, so it is not just limited to the physical brain. It bridges both worlds. Neural processes occur throughout the body and beyond it in the world around—this is known as haptic cognition, embodied cognition, or distributed cognition in Western science. This means that thinking and learning also occur outside of the brain in the objects and beings we interact with and the relationships in between. Cognitive science has so far found this only in humans, monkeys, and otters (although I have observed it in many other species, including bowerbirds and crows).

At the simplest level, when we hold a tool, our brain recognizes it as an extension of our arm. It isn't really part of our body, but it becomes an embodied extension of our neural processes. At more complex levels, the meaning we make with places, people, and objects and the way we organize interactions between these things become an extension of our thinking. Through meaning-making, we effectively store information outside our brains, in objects, places, and relationships with others. This is how spirit works.

If you use a familiar object to help you encode new knowledge that you are learning, then when you pick up that object or even just visualize it, you instantly remember what you learned. It has become a tangible metaphor, an overlap between the two worlds. This is why a lot of cultural objects have special significance in Aboriginal societies—knowledge is encoded into them in a creation process that is sacred. This is how traditional message sticks work. This kind of haptic knowledge is also encoded in relationships, which is why kinship systems are so central to our cultures. If you learn something with somebody, you might have trouble remembering it on your own but recall it in vivid detail when you are with her again, or if you think of her or call out her name.

There is a similar haptic relationship with Country, or with the Ancestors you might call out to when walking in particular places. Memories attached to places can be evoked by revisiting those places or even imagining walking there again. Haptic cognition also occurs throughout your entire body. There is knowledge and intelligence in your hands, feet, and even hair. Using your body consciously and meaningfully can unlock this intelligence. This is why any training that incorporates kinesthetic learning is so effective.

The most intelligent part of your body is actually separate from your central nervous system. Your gut has its own independent nervous system that is still a mystery to modern science. If your head is cut off, your gut will continue to function on its own until it dies from lack of oxygen. There is no name for this in English, but every Aboriginal language has a term for it. In my language it is *ngangk pi'an*. In Western Australia some people call it *ngarlu*. This is the seat of your big spirit, your higher intelligence. In Anglo cultures it is vaguely acknowledged when people refer to "gut instinct." In Eastern cultures it is the center of a person's chi. In all cultures it grounds a person in the living world and connects us to all things. In the Aboriginal world, the energy of the gut must be kept clear and constantly moving through mental, spiritual, and physical cultural activity, or it will become stagnant and make a person sick.

Although the gut is not connected directly to the brain (only indirectly through the cranial nerve complex), there must be interaction between these two systems in the same way that there must be interaction between the physical and spiritual worlds in the Turnaround, in order to maintain the whole system in a healthy balance. The interaction between the gut and the brain cannot happen mechanically through purely biological function, so it must be done through cultural practice. It is done by constantly making meaning in the worlds around us and within, transferring knowledge from one domain to another through use of metaphor. It is about making connections between things that would otherwise remain unconnected, using metaphors that are nonliteral and often seemingly irrational. This gives rise to both complexity and clarity.

Paradoxically, the more complex the meaning-making is, the clearer your thinking will be and the more likely you will be to remember new knowledge. Two test groups presented with the same list of long, unfamiliar words will have different results on a spelling test of those words depending on the complexity of the meaning-making they engage in. The control group is asked to memorize the spelling. The second group is asked to look up the meaning of the words. Every time, the second group does better on the spelling test. Any knowledge passed on as discrete information or skills is doomed to failure through disconnection and simplicity. Knowledge transmission must connect both abstract knowledge and concrete application through meaningful metaphors in order to be effective. Without this sacred and joyful act of creation, our systems become unstable and deteriorate. You need cultural metaphors of integrity to do it properly.

Working with grounded, complex metaphors that have integrity is the difference between decoration and art, tunes and music, commercialized fetishes and authentic cultural practice. When metaphors have integrity, they are multi-layered, with complex levels that may be accessed by people who have prerequisite understandings.

Working with metaphors is a point of common ground between Aboriginal and non-Aboriginal knowledge systems. We have a long tradition in Aboriginal society of ritual training in the use of metaphor during initiation into higher stages of knowledge. This is because metaphor is the way Law, Lore, Women's and Men's Business, ritual, Ceremony, and magic are worked ("ritual" and "magic" aren't capitalized here because anybody can do it). Powerful metaphors create the frameworks for powerful transformation processes, but only if they have that integrity. A metaphor that lacks integrity only damages connectedness—an action that is known as a curse in Aboriginal culture.

For example, I once visited an Aboriginal community school in the Northern Territory that was using the metaphor of Aboriginal fishing nets as an education framework. This may have worked as an idea of school and community weaving their different threads together to make the nets, then the students using the nets to catch fish, with the fish representing knowledge and social/cultural capital. But this was not the case. The fish in the net represented the children themselves, and the river represented the community, promoting a very problematic image of the school as an entity that captures children and takes them away to be consumed.

We have to be careful of the metaphors we use to make meaning, because metaphors are the language of spirit, and that's how we operate in our fields of existence either to increase or decrease connectedness within creation. We are the custodians who are uniquely gifted to do this work, so we need to do it consciously and with mastery, within cultural frameworks aligned with the patterns of creation. If we allow the I-am-greater-than deception to enter this process, all is lost.

So I must reflect now on the story I have used to introduce the knowledge I have carved painstakingly for months into my pointy parrying shield. Is that ghost story an appropriate metaphor to bring people into this knowledge of spirit? Is it funny enough to spark learning, or is it just silly? Is it also serious and interesting enough to spark deep concentration and engagement? Is it really a message that has been placed in my path that I am supposed to pass on in this yarn? Or am I just running all over cultural spaces like a mad emu and making a mess of things to feel better about myself?

Perhaps I was cracked on the head as a warning not to share any of this. I'm nervous about sharing these stories, but the crows outside my window are just relaxing and playing, not staring or screaming at me. Oldman Juma keeps telling me to share this knowledge without fear, but that hasn't always served me well. Maybe it's not meant to serve me at all but serve something else. An old Islander fella once told me to get my eye off myself, share freely, and it will all be taken care of. I am still struggling with that, even though I know he was right.

I guess I'll see what happens tonight. If that ghost comes back to torment me, I'll come back to delete things from these pages and only include what I put on the shield. If he doesn't, I'll just leave it as it is.

○

I wasn't troubled after that, so I left it as it was. The staff at Varuna followed up with me later to let me know they were bringing in a local Elder to smoke the old house out and put the spirits to rest at last. I can't tell you how happy this made me feel. How wonderful would it be if everybody could work together to bring about the same kind of closure and healing all over this world? Ghosts tend to linger where there is unfinished business and hard truths buried in shallow graves.

Advanced and Fair

This is a yarn about yarns—a meta-yarn, if you like. Us-two will find marginal stories and grand narratives twisted together in macabre ways here, as we stand in a noisy hall outside a high-school science lab. In front of us we see a fairly typical science display in a glass cabinet—a series of skull models representing human evolution. The skulls are colored in a sequence from most primitive to most advanced—from black to brown to beige to white. This same tone-chart sequence is repeated in the evolution posters on the wall, which clearly show dark apes progressing through eras of gradual melanin loss until finally arriving at the pinnacle of progress, the Nordic male.

As we stand there, us-two, giggling like naughty bubs at the big dirty joke in front of us, a teacher approaches, shaking his head. He has been unimpressed over the last two days by the professional development training we have been delivering to the school staff about Aboriginal perspectives.

"There you are," he says. "What are you looking at that for? I suppose you're coming up with another fuckin' conspiracy theory to make us all feel like racists then?"

"Yes," we say. "That's exactly what we're fucking doing."

My second son was born while I was writing this story. His name is Diver, after his great-great-grandfather, who was given that name by police when he was stolen from the bush as a child and made a ward of the state. He had no English name, so they called him that after he attempted to escape by diving into the river. His family and tribal affiliation are still unknown, however. This is a common ancestral story for a lot of Aboriginal people in Australia. Many are told they have no right to exist because they are unable to claim a tribal identity based on registered bloodlines or are unable to name these because of relatives who have passed for white and (understandably) wish to keep their prosperity intact. There are a lot of Indigenous stories like this, and if they don't line up with the larger narratives driving policies and funding, they are often burned up and blown away, along with the people who carry them.

One such larger narrative is that of the dying black race. This drove a policy of capturing and removing Aboriginal people in openly declared attempts to eradicate Aboriginality by breeding out dark skin in hopes of arriving at a "final solution." Well, it didn't work, although the narrative has enjoyed something of a revival lately. We're still here, and you can ridicule or deny or regulate or modify or limit our identities based on whatever new narratives you like, but we're not going away. Maybe at the start of the next century, some red-headed Chinese blackfella will be camping on what used to be Parliament House, cooking up a wombat and dancing up stories of how this all came to pass. And my boy Diver, blond as his mother's Irish dad, will still be here, telling our stories and passing on our culture, a proud scion of two strong Aboriginal families and a bunch of mad Celts.

That's little Diver's second name: Scion. His third name is Juma, so he is namesake of the old fella who keeps the forever yarns.

Our family stories will outlast the stories of this civilization, but at the moment they are almost invisible in the shadow of monolithic grand narratives like "progress." The narrative of progress is grounded in the myth of primitivism—the widely held assumption that life before the industrial era was brief, brutish, savage, and simple. This is contrasted with the myth of development, of advanced societies and people from Europe representing progress and enlightenment. There must be an upward trend to show, to keep the illusion alive. When the masses come down with *Avatar* depression or begin to chafe at the bleakness of their lives, they are reminded nightly on their screens of how terrible things used to be, how much better and longer and healthier our existences are today.

Unpacking these narratives and retelling them from our Indigenous perspectives highlights a few inconsistencies in the story and a few parts that have been left out. Our standpoint strips away the dogma that constrains our minds and potential, allowing some room for higher-order thinking that gives rise to intriguing questions.

For example, if Paleolithic lifestyles were so basic and primitive, how did humans evolve with trillions of potential neural connections in the brain, of which we now use only a small fraction? What kinds of sophisticated lifestyles would be needed to evolve such a massive brain over hundreds of thousands of years? What kind of nutritional abundance would be needed to develop such an organ, made mostly of fat? How does the narrative of harsh survival in a hostile landscape align with this fact? If our prehistoric lives were so violent, hard, and savage, how could we have evolved to have such soft skin, limited strength, and delicate parts?

The stereotypical caveman does not stand up well to this line of questioning, but still he persists in our hearts and minds, internalized by Indigenous and non-Indigenous people alike. It is such a prolific image in contemporary civilization that if I ask you to stop reading and picture it in your mind for a moment, you will then open your eyes and see that my next sentence describes that image perfectly:

A broad-nosed, thick-browed, hairy savage dressed in animal skins is standing with dull eyes and a wooden club in his hand, possibly over the supine form of a woman he just knocked out to drag back to his cave.

For this chapter I paid homage to that Western cliché by carving two wooden clubs. In my clan these are referred to as *yuk puuyngk*, or Law sticks. They communicate custodial claims to places and stories, asserting cultural authority and even ownership of objects and resources. They can be stuck into the ground to show a claim to a place or to objects alongside them. The stories of the person carrying them may be etched or painted on the business end of the club. Smaller ones may also be used as message sticks to carry stories and knowledge to different groups and territories. But if you dug them up in the ancient dirt at the bottom of a cave, it would be hard to imagine any utility for them beyond bludgeoning.

The two clubs are carved from beefwood, and they are flawed with a number of cracks in the surface of the wood that have been filled with a black sap used as a kind of glue or putty. There are cracks in the stories I'm sharing here too, the grand narratives and the cheeky counter-narratives I am presenting to disrupt them. I yarned with a lot of different people while I carved out the ideas and stories in this chapter, but nobody would engage with my critique of primitivism. Their eyes just glazed over, and I felt like I'd hit some kind of firewall installed through Sunday-morning cartoons. They all engaged with my story about Prussia though, which surprised me because it's probably the most lunatic, incendiary rant that has ever fallen from my lips. But before we get to that, we'll continue the lonely business of tipping the sacred cows of primitivism and progress.

The myth of the primitive is a constant touchstone and counterpoint to the narrative of progress. It is used across science, with almost every explanation of human phenomena containing the words "back when we were cavemen," with pop-science pundits using this mythology to justify everything from social media addiction to rape. My favorite assertion was from a doctor who said the first medical tool was a stick that cavemen used to poke unmoving people to see if they were dead. Which suggests they are able to determine a hell of a lot of cultural information from a small pile of bones discovered around the world, the sum total of which would fit in the back seat of your car.

On the bright side, researchers are always happy to modify this story when new information comes to light. For example, when DNA from the iconic Neanderthal was discovered in all Europeans, overnight announcements changed the narrative. *Hey, we're discovering that Neanderthals were actually a lot more sophisticated than previously thought!* Not every new discovery entered the story though. When it was found that Neanderthal women carried much the same suite of bone injuries as men (similar to the fracture patterns of rodeo clowns), there was a brief silence before the "men were hunters and women were gatherers" narrative continued unchallenged.

The clean healing of those injuries also suggested that people cared for each other in ancient times and had fairly advanced medical practices as well as a diet rich enough to mend bones quickly. There was even evidence of disabled people being cared for well into their twilight years. But that doesn't fit the dominant narrative of brutish misery and short lives, with only the strongest and most psychotic individuals dominating and surviving against the odds, killing everyone else and spreading their superior seed.

There is a reason not many ancient human bones have been discovered. Until very recently in human history, most people had funerary practices that included sky burials. This involved placing the body on a platform to be eaten by birds and animals, with remaining large bones carried by mourners for a respectable period before being stored in trees or caves or logs or dunes, or broken up into small pieces as part of mourning rituals. Those big bones were usually the skull and thigh bones. Later in history those bones would become a symbol of heathen terrorists known as pirates. Still later, they would be used as a symbol warning of poison and death. So the sky burial has been forgotten, and people are left wondering why there are not many bones to find, and why the ones they do find are fairly miserable specimens. It is probably because those specimens belonged to rogues, bandits, and outcasts who had nobody to care for them after death—which is an interesting data set to build an entire prehistory from.

But still, they built that narrative. You'll find no other area of science (outside of pro-smoking doctors in the 1950s) that has manufactured more fanciful tales from limited evidence. These tales were projected onto all encounters with southern hemisphere Indigenous people in an orgy of confirmation bias that manifested as a weirdly diligent mass hysteria. Stories are powerful tools and can be even more powerful weapons in the hands of malignant narcissists. If you want to take control of your life or work toward some kind of sustainable change in the world, you need to harness the power of story.

This is a symbol I use to mark a way of thinking I call story-mind. There is more to narrative than simply telling our stories. We have to compare our stories with the stories of others to seek greater understanding about our reality. It is a test of validity and rigor for new knowledge. The symbol shows two people sitting, bringing their stories together to share through sand talk to extend their knowledge. False narratives do not stand up well to this kind of analysis. Story-mind is a way of thinking that encourages dialogue about history from different perspectives, as well as the raw learning power of narrative itself.

Narrative is the most powerful mechanism for memory. While isolated facts go only to short-term memory, or to midterm memory with repetition (as with study for exams), story goes immediately to long-term memory. If you can make up a story connecting metaphors, locations, and language triggers to help you remember something you are studying, it will save time and increase your long-term recall.

Stories are also called yarns, but "yarning" as a verb is a different process altogether. The symbol represents the yarning process as well as narrative, because this is the process by which stories come together and begin to have meaning. Without yarning, stories are just something to put children to sleep. There has to be an exchange of stories if you want to be awake and grown.

Yarning is more than just a story or conversation in Aboriginal culture—it is a structured cultural activity that is recognized even in research circles as a valid and rigorous methodology for knowledge production, inquiry, and transmission. It is a ritual that incorporates elements such as story, humor, gesture, and mimicry for consensus-building, meaning-making, and innovation. It references places and relationships and is highly contextualized in the local world-views of those yarning.

It has protocols of active listening, mutual respect, and building on what others have said rather than openly contradicting them or debating their ideas. There is no firm protocol of only one person speaking at a time, although the mutual respect protocol ensures that interjections are in support of what a speaker is saying, enriching what is being said. There is no "talking stick" protocol. (The talking stick idea was appropriated by the West from Native American culture.)

This back-and-forth yarning style neutralizes the unpleasant phenomenon that occurs in many conversations, meetings, and dialogues of one person grandstanding and waffling on while the rest of the group drowns in polite boredom. (Monologues are rare in Aboriginal culture unless a senior person is telling a long story or an angry person is airing grievances.) There is a lot of overlapping speech that makes yarning vibrant and dynamic and deeply stimulating. It is nonlinear, branching off into diverse themes and topics but often returning to revisit ideas in ways that find connections and correlations between diverse sets of data that would otherwise not be found in more analytical modes of dialogue. There may be periods of comfortable and communicative silence that are reflective and not considered to be awkward. The end point of a yarn is a set of understandings, values, and directions shared by all members of the group in a loose consensus that is inclusive of diverse points of view.

The primary mode of communication in yarns is narrative—the sharing of anecdotes, stories, and experiences from the lived reality of the participants. Sand talk may be incorporated as people sketch images on the ground (or even in the air) to illustrate a point or to map out a place. Physical demonstrations are included as people act out events occurring in stories. Sharing drink or food is often part of the ritual—most commonly cups of tea today. Often yarning will occur around a shared material cultural activity like weaving, painting, string-making, Ceremony preparation, and even things like crossword puzzles and setting up birthday party decorations.

Yarners will usually sit in a group so that everybody can see everybody else, or in a rough circle. Whatever shape the formation takes, it has no stage or audience spaces. This removes hierarchical barriers to consensus and also avoids the shame that often comes with being in the spotlight or having to speak in front of an audience. Some group members may have more authority and respect than others if they are senior people, but that authority is usually used to pull people back into line when they break protocol, rather than to pursue ego-driven agendas. *Usually.*

Yarns and yarning can be more than tools to enhance memory and engagement. They can be a disruptive innovation that is empowering and liberating when transferred to contemporary media. Unfortunately, this potential is usually lost in the widespread Indigenized genres of dreamtime stories and personal recounts. While this kind of narrative is important, it lacks the power of a critical Indigenous lens on the world. "Strong Indigenous voices" need to be doing more than recounting our subjective experiences; we also need to be examining the narratives of the occupying culture and challenging them with counter-narratives. This is difficult though, because every time we win some small space for Indigenous critical perspectives in any institution, conservative pundits insist we should be promoting the history of Western civilization, particularly in the education system; and I agree—we should examine the history of this civilization in schools, very closely.

There is a reason ideological battles and culture wars filled with rhetoric about patriotism and nation-building are fought around schools and schooling. Schools are sites of political struggle in this civilization because they are the main vehicles for establishing the grand narratives needed to make progress possible. The entire history of globalization hinges on the story of modern public education, how it began, and why. I often wonder what would change if people were able to see this story retold from the perspective of an Aboriginal person reading back through old federation documents and the earliest syllabuses from Prussia.

The answer is: it would be an outlandish conspiracy theory that has no place in a glass cabinet full of skulls. That's why I'm going to tell it here.

While most of the facts are verifiable, I have been very selective in which facts I used to build the narrative. I created the story to illuminate the way history can be twisted to suit the interests and narratives of the people who write it. But mostly I wrote it for a laugh—it is fun to imagine what history would look like if it were written not by the winners but by losers like me.

The story of modern public education, then, is a story of transition between an age of imperialism and an age of modern globalization. It begins, like all stories about civilizations, with the theft of land from Indigenous people. The people were the Prusai, natives of an area between modern-day Germany and Russia, who lived there from at least 9000 BCE. They traded amber and hemp across Europe and into Asia, but mostly they lived by hunting and fishing. They maintained this society right up until the thirteenth century.

In the south, trouble had been brewing for centuries. Germanic and other Nordic refugees (refugees from previous Roman invasions, from rising sea levels, and from starvation as a result of degraded soil caused by recent incursions of agriculture) went viking across Europe. "Viking" was a verb meaning "raiding" in those days, and *these* boat people really were a problem. They had overrun Britain and changed that island forever, although Roman and Celtic invaders had already been there before them, so the poor old British copped a triple dose of colonial abuse. The Prusai, however, were lucky enough to escape the worst of this colonization process for many centuries, and—for a while, at least—they continued their traditional lifestyle, along with many other Indigenous nations to the north.

There weren't single big nations like today but plurinational groups of regions: lots of regions all with different laws, languages, and customs—very much like Australia was before colonial occupation. Big countries with one law, one language, and one people are a very recent invention designed to facilitate more effective control of populations and resources for economic purposes. This is why, after the Romans had left the Germanic regions, the rich landowners struggling for dominance there worked hard to restore the Roman system of social control. They fought to reinstate this power system for a thousand years, with many small states battling each other for supremacy, the Roman eagle standard emblazoned on their coats of arms. This obsession with Rome would cause some problems down the track, particularly for the Indigenous Prusai to the north.

In the thirteenth century, an organization called the Order of Teutonic Knights broke away from the other German regional groups and decided to create its own new state. The site the knights chose was Prusailand, so they invaded and exterminated or assimilated the Prusai people, making the entire population Teutonic. In classical and fantasy art, however, you won't find any images of knights wiping out Indigenous people. What you will find instead are armored heroes bravely slaughtering beasts, dragons, and mythical monsters. These creatures came to represent the tribal cultures of the world: the romantic European image of the knight slaying the dragon is actually a hidden reference to the systematic genocide of what were called pagan peoples. This European tradition of propaganda in which victims of genocide are portrayed as dangerous animals was later used to great effect against the Jews and even our own "mob" here in Australia, who up until half a century ago were often considered animals rather than human citizens.

By 1281, the Order of Teutonic Knights had all but wiped out the native Prusai and created the new state of Prussia. The interesting thing is that these "white knights" had been heavily involved with the Crusades, in which the Roman Church had been fighting a Christian war for centuries to take over Jerusalem and other holy places. They failed so badly that instead of bringing Europe to the Middle East, they brought the Middle East back to Europe in the form of a system of government that they had seen there and liked. This system was in its final stages of decline during the Crusades, with most of the Middle Eastern forests and farming land stripped bare and turned into desert by the ravages of the world's first civilizations. In the following centuries, the survivors had begun returning to more sustainable ways of life—tribalism, subsistence agriculture, and pastoralism (a way of life that would later be turned upside down again by twentieth-century Anglo oil interests).

The failed model borrowed by the Teutonic Knights wasn't invented in the Middle East. It had its origins in an unsuccessful Asian experiment of large states with total government authority and rampant expansion and production. This was completely alien in Europe, which was used to a system of petty warlords and oligarchs struggling chaotically over dwindling natural resources, while local peasants in villages persisted much as they had since the beginning of the Iron Age, periodically disrupted by the activities of the powerful. The exotic new system introduced by the Teutonic Knights was all about absolute power concentrated into one highly organized central government that would control the daily lives of all.

Remember too that these new Prussians had just spent a thousand years trying to replicate the system of control that they had experienced under the Romans, who had originally conquered Germania. (Britain and the US later mastered Rome's imperial method: a system of establishing Indigenous elites to keep conquered peoples in check, promoting lateral violence and competition to make subjugated peoples self-policing vassals.) Prussia even adopted the Roman symbol of the eagle as a logo, which was later picked up by the United States and the Nazis. Rome introduced mesmerizing dreams of power and control that have not been easy to shake, even in modern history.

By the eighteenth century, Prussia, under Frederick the Great, had become one of the greatest powers in Europe, despite its small size and lack of natural resources. This was due to the fact that it had a larger permanent military force than anyone else. No other country could force so many of its citizens into the army full-time. The Prussian system was one of total control, which successfully managed to coerce the population into complete submission to the will of the government. Creating a massive standing army was not a problem for them. (Over a century later, the US military would adopt their formula for maintaining permanent standing armies on the advice of a Prussian military consultant named Friedrich Wilhelm von Steuben.)

Prussia didn't stop there. The more rights it stripped from Prussian citizens, the more powerful it became. Frederick the Great's nephew continued this process, depriving every adult of all rights and privileges.

Then in 1806 the Prussians suffered a shattering military defeat at the hands of Napoleon. After their beaten soldiers fled from certain death, the Prussians decided to turn their attention to the children, realizing they had to start young if they wanted to instill the kind of obedience that would override the fear of death itself.

The government decided that if it could force people to remain children for a few extra years, then it could retard social, emotional, and intellectual development and control them more easily. This was the point in history when "adolescence" was invented—a method of slowing the transition from childhood to adulthood, so that it would take years rather than, for example, the months it takes in Indigenous rites of passage.

This delayed transition, intended to create a permanent state of childlike compliance in adults, was developed from farming techniques used to break horses and to domesticate animals. Bear in mind that the original domestication of animals involved the mutation of wild species into an infantilized form with a smaller brain and an inability to adapt or solve problems. To domesticate an animal in this way you must:

1. Separate the young from their parents in the daylight hours.
2. Confine them in an enclosed space with limited stimulation or access to natural habitat.
3. Use rewards and punishments to force them to comply with purposeless tasks.

Effectively, the Prussians created a system using the same techniques to manufacture adolescence and thus domesticate their people.

The system they invented in the early nineteenth century to administer this change was public education: the radical innovation of universal primary schooling, followed by streaming into trade, professional, and leadership education. It was all arbitrated by a rigorous examination system (on top of the usual considerations of money and class). The vast majority of Prussian students (over 90 percent) attended the

Volksschule, where they learned a simple version of history, religion, manners, and obedience and were drilled endlessly in basic literacy and numeracy. Discipline was paramount; boredom was weaponized and deployed to lobotomize the population.

This system worked so well that Prussia became one of the most powerful countries in the world at a time when the idea of "nations" (rather than regions, kingdoms, tribes, or city-states) was first being promoted as the dominant form of social organization on the planet. The Prussians began to make plans to spread the institution of schooling as a tool for social control throughout the world, as it facilitated the kind of uniformity and compliance that was needed to make the model of nationhood work. The US could testify to the effectiveness of Prussian education as a tool for domination and power, as American educators had been making pilgrimages to Germany for more than half a century. Excitingly, test schools across America proved that the artificially induced phenomenon of adolescence was achievable outside of Prussia, too.

In 1870, Prussia got its revenge on France by annihilating the French military in the Franco-Prussian War and immediately established Germany as a unified nation-state—the dream of the Teutonic Knights finally realized. After that, the Prussian education system (and the new institution of extended childhood) became all the rage around the Western world. It was modified to some extent, probably because the Prussian model seemed a bit weird, even to the power-hungry ultra-rich of Europe—it was so all-encompassing that women were required to register each month with the police when their menstruation started. Prussia was described jokingly as an "army with a country" or a "gigantic penal institution." Towns and cities were built like prison blocks, gray grids of rigid cubes and plain surfaces. The government worked hard to "cleanse" the society of homeless people, gypsies, Jews, and homosexuals as they expanded and enforced their embryonic doctrine of eugenics. Their motto for education was *Arbeit macht frei* (Work sets you free), a slogan that the Nazis adopted and later placed above the gates of concentration camps, including Auschwitz, used for Jewish slave labor and extermination. There are many schools in Australia today with a similar motto in Latin: *Labor omnia vincit* (Work conquers all). Now, as ever, the creation of a workforce to serve the national economy is the openly stated main goal of public education. And, as ever, the inmates of this system are told that their enthusiastic compliance with forced labor will be in their best interests at some future point.

Germany's compulsory education system expressed six outcomes in its original syllabus documents:

1. Obedient soldiers to the army.
2. Obedient workers for mines, factories, and farms.
3. Well-subordinated civil servants.
4. Well-subordinated clerks for industry.
5. Citizens who thought alike on most issues.
6. National uniformity in thought, word, and deed.

And it spread like wildfire: to Hungary in 1868, Austria in 1869, Switzerland in 1874, Italy in 1877, Holland in 1878, Belgium in 1879, Britain in 1880, and France in 1882. From there it quickly expanded further to European colonies, including Australia.

As we've seen, the US had been involved much earlier, with even Benjamin Franklin advocating the Prussian system. In 1913 Woodrow Wilson established the Federal Reserve, copying Germany's centralized banking system too: this way, the state would control both learning and money, just like Germany did.

As the twentieth century wore on, more interesting links emerged between Germany and the US, both drawing on the symbols and dreamings of ancient Rome—because Germany's old obsession with ancient Rome hadn't gone away. They called their leader "Kaiser," German for "Caesar"; they adopted the symbol of Roman fasces, bundles of rods with an ax that once represented Roman state power. The US followed suit. American education documents emerged with those same symbols printed on the covers, and today the fasces are still a prominent symbol of American power, proudly on display in many official ceremonies.

The Roman fasces came to represent a whole modern belief system around social control and national domination—that's where fascism got its name—and a version of the Roman salute was famously adopted by the Nazis.

In that period, when Hitler was *Time*'s Man of the Year in America, the pseudoscience of eugenics that the Nazis so enthusiastically adopted was popular throughout the Western world. It purported to legitimate a decades-old tradition of white supremacy that had earlier informed the nationalist values established during Australian federation and exemplified by the White Australia Policy.

Not just in Australia but all around the world, new systems of education, nationalism, finance, corporatism, and social control were informed by fascist ideas and theories from Germany and the United States, encouraging the extermination of Indigenous people and minorities, just like the white knights of yore. Cataclysms followed as new nations that had missed out on the empire-building activities of the age of discovery tried to catch up with their land-rich neighbors. When the smoke cleared, lands and power and blame were redistributed unevenly among the survivors, and a new world emerged with new stories providing a sanitized history of good triumphing over evil. In Italy, for example, it used to be common knowledge but is now all but forgotten that Hitler's fascist partner in crime, Mussolini, exterminated the Cavernicoli, a cave-dwelling people who were still maintaining a Paleolithic culture.

But the structural racism installed through Prussian-style schooling and the eugenics movement would not be discarded, merely rebranded. Later, following long civil-rights struggles and campaigns for social justice, racial inferiority was renamed "cultural difference." Racial integration was called "reconciliation." In the colonies, assimilation was relaunched as "Closing the Gap." The language became more politically correct, but the globalizing goals of cultural uniformity, economic compliance, and homogenized identities remained the same.

In my crackpot version of this history, public schooling plays a principal role in the story of transition from one age to the next. It is by no means a complete account, but I hope this marginal perspective is far enough "out of the box" to provoke some questions regarding the sustainability of the global systems that shape our minds and lives. Where is our current turbulent period of transition taking us? Do we want to go there? What form will knowledge transmission (aka education) need to take during this transition? Us-two may also tentatively wonder whether our minds are now too domesticated and shriveled even to contemplate these questions effectively.

A good way to begin might be to listen to as many divergent versions of this history as possible, from many points of view. I have particularly learned a lot from feminist versions of this story, but also from libertarians and poets and neo-conservative professors and the alt-right and incarcerated people. Most important, I have learned from yarns with children, the inmates of the education system.

Suddenly I think of us-two standing back there at the high-school science lab, laughing at the glass cabinet with all those evolution models, and I wonder what kinds of stories we will share and seek together from now on. Shhh, the teacher's coming!

"There you are," he says. "What are you looking at that for? I suppose you're coming up with another fuckin' conspiracy theory to make us all feel like racists then?"

"No, not at all," says us-two. "We're just admiring your lovely skulls. Tell us the story of how they came to be."

Romancing the Stone Age

I squirm at first
Can't listen if I can't believe
These claims made of Adam and Eve
From Oldman there sitting—departure gate
Red dirt etched biotic under fingernails tapping keys
Dust like powdered blood from his Eden in the far far north
Voice file rasping dried like crimson ironwood tree sap and
smoke
Coaxed from those same leaves to warm this place
exorcising bad spirits
Knowledge billowing thicker and greater with each new
stage of story spiraling
Suddenly smothered
With my Paleo-Luddite doubt
About Moses coded into Uluru and
Departure gate signage with hidden meanings and
Dreamings

In lettered settler protocol messaging patterning
Ancestral missives in the food court?
Everything Dreaming he asserts annoyed
Everything even dumplings and luggage and bustle
Still pattern still sign still a something
Whirlwind and willy wagtail and www dot all one
Our old people whispering through the air-con
Venting plenty big story if you can listen
Ceremony.com
Flash mobs dancing not hearing the
Tree behind the coal behind the steam behind the
    charger
The bauxite in the beer can
The sand in the bottle
Rare earth metal circuitry conducting spirit real time
From deep time this synthetic way a glitch
Walls of fire and a world burnt to
Reboot a shadow of the connectivity that was
While pirates board the mother
Cane toad dreaming uploads a toxin sponge
Radiation halving this half life
Red eye and amphibious claw mutations
Bringing back to land and waters every particle
Like all the families of the world coming back to that rock
First place first people blockchaining ways
Of return and okay Jesus call him if you like
Jumping up with one foot off those songlines
Powerlines
And a baby boy adrift on a river
Let my people come

Seven spirit families seven sisters stars
Named the same way across this planet every tribe
Always pursued and the little one left behind
But her return flight is booked for that special day
And it's here today because all those tribes are gathered
  from exile
At gate 13 all those global bloods one time and place again
For the next big bang boom echidna smashing turtle
  hexagonal fractal
True god next iteration creation event
And I doubt it and he shames me shows me
Shouts the last word of the story
POWER!
And all the lights go out white noise stops and screens go
  blank
Terminal shut down with one word
And twin tornadoes dance over the horizon
See?
Those two serpents
First man first woman story
That baby boy return running on one foot next Thursday
Thor's day making storm and lightning see?
Mungo Bethlehem Reykjavík same way smart city
Exodus and return like breathe out breathe in
That inhale is just about to
FINISH!
And the lights flicker back on and the crowds sigh relief
Checking screens for numbers
Patterns
See?

True story, that little song there. Oldman Juma did indeed shut down an airport with a word and a handclap. He has a lot of power and knowledge, but he also says a lot of wild things that I just can't bring myself to believe, even when he proves that they're true. There's way too much biblical stuff mixed up in there for my comfort, and I just can't accept the idea that cane toads are our friends. You might be feeling the same way about some of the things I'm saying in this book, and I empathize with you. I get frustrated when I hear mind-blowing wisdom mixed up randomly with ideas that seem to follow no rational path of logic. But while Oldman Juma's wisdom may blow my mind, his seemingly irrational ideas expand it, forcing me to look at things from impossible angles. It still annoys the hell out of me though.

He looks at printed signs and sees them as ancestral signals the same way we might read omens in the movements of animals and plants and winds. "See?" he always asks. No, I don't! I don't see that this CAUTION sign means C-Cousin, A-Aunty, U-Uncle, and so on. It's a foreign language on a fabricated steel sheet! How is that my Dreaming? All Dreaming, he says, all language, all message, see? See?

He interacts with his phone like it's a benevolent alien message stick. I get frustrated because I know he's right and that I need to embrace these things as part of creation or I'll miss patterns I need to see. But I'm a Luddite, and it's not easy. I used to get wild when my own people would tell me I'm not a proper blackfella if I don't have a mobile phone. I resisted as long as I could. From 2010 onward, my employers insisted I carry office phones on the road for emergencies and meetings, but I seldom used them. I got my first ever personal mobile phone in 2016 and then watched my brain fall apart like wet cake over the next few months.

I can't imagine what a flashy phone with all the bells and whistles would do to me. I can only get prepaid phones because, like most people in my community, my credit rating is not good enough to get a phone plan. My phone is cheap and hasn't got much memory so I can only fit a handful of apps on there, but those few apps have been enough to completely change the way my mind and body are wired. I feel numb after I've been using it, like when your arm falls asleep. No wonder people have become so weird over the last couple of decades. I understand now.

I made a rounded parrying shield out of *thaanchal* wood for this chapter, but it isn't very good. It doesn't look right, so I keep it face down and ignore it until I need to pick it up and remember what I'm writing about. I wasn't fully focused when I made it because I was brain dead from trying to make the email work on my phone. I don't know what the IP address is! What the hell *is* an IP address?

But I need to embrace this technology now because I know Oldman is right. I've been yarning with a lot of people and reading about it, enough to see how it's shaping my life and will radically change the world in the next few years. I'm not looking forward to that at all. I know it's part of creation and there's no valid way to separate the natural from the synthetic, the digital from the ecological. Maybe I'm just clinging to a romanticized vision of a Paleolithic past that's just as prejudicial as the myths of primitivism I criticize. There are as many fads and movements putting cavemen on a pedestal as there are myths that denigrate them, and perhaps that woo-woo stuff is equally damaging. Paleo-dietitians, new agers, and anarcho-primitivists have marketed all kinds of ideologies in recent years. Maybe I need to examine some of these more closely to put things in perspective.

So for this chapter I yarned with the most respected and widely published anarcho-primitivist thinker I could find, John Zerzan. But first, I'd like to address this growing cognitive numbness I'm experiencing by grounding us in an Indigenous way of thinking that might insulate us against it, so we can listen properly to what Zerzan is saying.

This symbol represents either a wooden dish (coolamon) or a wooden shield. There are two interpretations, to connect with both women's and men's cultural activity. The three lines represent pathways to high levels of knowledge and engagement that can be attained through the intense states of concentration required to make and use these cultural items. They may be used in diverse activities such as dancing and food preparation, requiring a deep connection to spirit and the boundless energy and discipline involved in Aboriginal cultural activity. Muscular memory and even cellular memory (inherited memory) may be activated in these tasks.

The demands of the clock become irrelevant as true "blackfella time" kicks in and we become immersed in the activity. We may attain similarly heightened mental states in sleep (or even in trance) as our minds organize terabytes of information in dreams. Even walking can produce this trancelike state of deep thought—often walking and yarning for extended periods with others stimulates high levels of learning and memory because of this mental state, which I call ancestor-mind.

The shield reminds us to protect our knowledge and our minds from the white noise of civilization, that constant background hum of machinery that has been found to cause brain damage and lower IQ. On another level, there is cultural white noise as well that is damaging to knowledge and ways of living. Engaging with this ancestor-mind way of thinking is a protective practice that reverses this damage. The coolamon may be used to hold an infant, ocher, or nourishing food. It represents the holding of knowledge and the nurturing of our children's minds, undomesticated minds that are exactly the same as the minds of our Ancestors, uncolonized and vibrantly connected to the worlds around us.

Anybody who has small children, or works with them, will be familiar with the qualities of an undomesticated mind. It is wild and unschooled, teeming with innate knowledge processes. Children perform tasks they have not encountered before and you wonder, "Where the hell did they learn that?" They play with absolute dedication and fierce concentration. They learn languages perfectly, to the limits of their adult role models, without explicit instruction and at a phenomenal rate. Most of what we learn in our lifetimes today is during the first few years of childhood. This explosive period of learning ends, in industrialized communities, around the age of eight, when the highly focused attention of play and learning is schooled out of the child and redirected into less stimulating educational activity.

After three or four years of schooling, the nucleus basalis, which forms sharp memories in the brain, falls into disuse and decays. This is the part of the brain that makes learning so effortless for small children, and it is always activated in undomesticated humans. But neuroplasticity research has shown that damage to the nucleus basalis can be reversed by reintroducing activities involving highly focused attention, which results in a massive increase in production of acetylcholine and dopamine. Using new skills under conditions of intense focus rewires billions of neural connections and reactivates the nucleus basalis. Loss of function in this part of the brain is not a natural stage of development—we are supposed to retain and even increase it throughout our lives. Until very recently in human history, we did.

Bearing this in mind, the reclaiming of Indigenous ritual and cultural activities as exercises in concentration, rather than just performances or soft-skill craftwork, may be just what is needed to grow or repair the minds required to create complex solutions for sustainability issues.

Every book I have read on memory and brain science has had some reference to the genius of childhood, and many books encourage people to learn and think like a child if they wish to increase their IQ or capacity for memory. I always wonder why we work so hard to train this genius out of children in the first place, rather than building on it.

Perhaps, on some level, we suspect the source of a child's genius lies not only in the mechanics and chemistry of the brain but in something far more powerful and difficult to explain. Perhaps these biological neural functions are not the source of mental activity, but a response to thoughts originating elsewhere. Could it be possible that we fear the true source and potential of our minds, the raw power of our Ancestors' cognition that we see reflected in the eyes of our children?

Some aspects of consciousness, knowledge, and knowledge transmission have not been explained or proven scientifically and are therefore avoided in cognitive science. I'm calling these aspects "extra-cognitive" for want of a better word. They include the messages that land and Ancestors bring to us—a bird or animal behaving strangely, a sudden wind gust, a coincidence that highlights a deep meaning or revelation, a burst of inspiration. These are the things that make knowledge processes sacred and magical.

How do animals know migration routes they have never individually experienced before? How does a woman intuitively recover a weaving technique that disappeared over a century ago? How do ideas and solutions appear to us in our sleep?

Perhaps in this case the "how" doesn't matter. Perhaps the mystery of it all is essential to how it works. Most people from most cultures are aware of this extra-cognitive activity in their lives, despite the lack of evidence to support its existence. Even people committed to rational and analytical ways of thinking and being will still refer to "gut instinct" or will find that a complex problem is better understood if a person "sleeps on it."

Inspiration is something that has been relegated to the arts rather than the sciences, although stories of "eureka moments" in scientific discovery are still celebrated. One fella sits under a tree and gets knocked on the head by an apple. Another fella takes a bath and has a revelation. But creativity is now widely regarded as a vaguely defined skill set falling randomly on individual geniuses. Deep engagement encompassing mind, body, heart, and spirit has been replaced by a dogged ethic of commitment to labor and enthusiastic compliance with discipline imposed by authority. While it may be proven that internal motivation is more productive than external pressure, the uncertain and unsettling sources of this inner power are threatening to hierarchies, so intrinsic control methods of organization are generally ignored in both education and the workplace. Or they are co-opted into "self-management" protocols that involve internalizing our administrators and doing the job of monitoring or managing for them—an arrangement not unlike the child who always has the voice of an abusive parent in his head.

Inspirational connections to unseen inner and outer worlds are a part of Aboriginal Knowledge transmission all over Australia. This connection is interwoven with every learning experience within the communities of First Peoples. It is ritual. It is the force that animates all Aboriginal Knowledge—a spirit of genius that shows the difference between yarning and conversation, Story and narrative, ritual and routine, civility and connectedness, information and knowledge. Most of all, it highlights the massive divide between engagement and compliance. Most of us today are living in a state of compliance with imposed roles and tasks rather than a heightened state of engagement. We are slaves to a work ethic that is unnatural and unnecessary.

The Protestant work ethic developed from the theories of a powerful clergy. Any time not spent at work or prayer was a chance for the Devil to enter a person and whisper evil suggestions, perhaps something like, "What the hell am I getting out of this?" The clergy conjured up all kinds of demons and evil spirits to frighten workers into compliance. My favorite was the Noonday Demon, a wicked entity that would enter a person if they considered having a siesta. Slave-like labor without lunch breaks would purify the soul in preparation for paradise. *You can take a holiday when you die!*

Us-two, we still endure longer work hours than our roles require today, for reasons of social control rather than productivity. It's difficult to find the mental space to question systems of power when we're working eight hours, then trying to lift heavy weights that don't need lifting or pedaling bikes that go nowhere for an hour so we don't die of a heart attack from being stuck for a third of our lives in a physically restrictive workspace. We sleep for another third of our lives (although not if we have small children), then the rest is divided between life-maintenance tasks, commuting, and using the few remaining minutes to connect with loved ones, if we still have any. Somewhere in there we also need to find time to study and retrain, unless we want to finish up homeless when our industries inevitably collapse or change direction.

The job is the unquestioned goal for all free citizens of the world—the ultimate public good. It is the clearly stated exit goal of all education and the only sanctioned reason for acquiring knowledge. But if we think about it for a moment, jobs are not what we want. We want shelter, food, strong relationships, a livable habitat, stimulating learning activity, and time to perform valued tasks in which we excel. I don't know of many jobs that will allow access to more than two or three of those things at a time, unless you have a particularly benevolent owner or employer.

I am often told that I should be grateful for the progress that Western civilization has brought to these shores. I am not. This life of work-or-die is not an improvement on preinvasion living, which involved only a few hours of work a day for shelter and sustenance, performing tasks that people do now for leisure activities on their yearly vacations: fishing, collecting plants, hunting, camping, and so forth. The rest of the day was for fun, strengthening relationships, ritual and ceremony, cultural expression, intellectual pursuits, and the expert crafting of exceptional objects. I know this is true because I have lived like this, even in this era when the land is only a pale shadow of the abundance that once was. We have been lied to about the "harsh survival" lifestyles of the past. There was nothing harsh about it. If it was so harsh—such a brutish, menial struggle for existence—then we would not have evolved to become the delicate, intelligent creatures that we are.

But there I go, romanticizing the past again. Time to yarn with somebody who's far better at that than I am, a famous thinker in the American anarcho-primitivist community whose work has often been dismissed as "noble savage stuff." John Zerzan is an American of Czech descent. His books include *Running on Emptiness, Twilight of the Machines*, and *Elements of Refusal*. He argues that civilizations oppress people through scarcity paradigms while Indigenous communities have free societies based on paradigms of abundance. He says we need to reform civilization by drawing on the knowledge and values of Indigenous societies.

John is unusually self-critical and reflective on the topic of romanticizing the past, admitting to me, "I have done a bit of it, such as in my book *Future Primitive*, where an overall picture was drawn that valorized hunter-gatherer life. Romanticizing the past is a danger when one generalizes." He also makes some good points defending his work against the "noble savage" slur.

"That's a rather typical postmodern dismissal of a thoughtless variety. I can't speak for what 'noble' means when applied to those outside civilization, but I think I damn well know what 'ignoble' is when it comes to the horror show that was set in motion from domestication onward. The Indigenous dimension is my touchstone, a source of inspiration."

We talk a lot about the nature of work in industrial and preindustrial cultures, and John suggests that the very concept of work is a modern one that did not exist as a separate activity in human society until recently. I think about this for a while and realize that there is no word for work in my home language and none in any other Aboriginal language I have seen.

John shares my unease about the "eat like a caveman!" branding and marketing of Paleo diet fads and the strange fascination these hold for people. "However," he adds, "Paleo diet popularity probably indicates that the yearning for an authentic life is not yet dead. But such gestures and fads don't go far at all on their own. A fundamental change from the dominant culture is needed, and it's a lot harder to be heard on that than on the topic of some fascinating caveman diet."

Nonetheless, John strikes a hopeful note: "The dominant order always fails. Every civilization has failed, and this global one is failing grandly, obviously. Our enemy has no answers. This makes me a bit hopeful. Change is possible because it is necessary." I hear some mild alarm bells toward the end there when he says "our enemy," followed by short, rapid sentences, and I wonder if I've now been placed on a CIA watch list somewhere for even having this yarn. But perhaps I'm already on one. (I used to send money to a Mohawk protest group blockading developers who were trying to build a golf course on their ancestral burial grounds. I later found they had been listed as a terrorist group under the Patriot Act.)

My yarns with John end on some intriguing thoughts about the changing nature of work, leading to reflections on the nature of memory and the power of our minds. It is possible that the resources to support technological growth will last for several years yet, certainly long enough for most jobs to become automated, potentially freeing up a lot of workers to pursue more satisfying and sustainable lives. The only problem is, what will they do for money when their jobs are gone? How will they eat? There will soon be a lot more irrelevant, disgruntled, hungry people in the world. This has never been a circumstance that has added to the longevity of a civilization. People don't tend to remember these lessons from history, though. There seems to be collective memory loss, even over the short term, in some of the adolescent cultures of the world.

The amazing thing discovered by brain researchers is that humans never actually lose memories—we just lose access to them, but they still somehow guide our actions. We might lose conscious access to the traumatic memory of our hair catching fire yet still remain afraid of flames, for example. Part of us remembers.

Each of us has indirect access to the memory of every moment of every experience of our lives. It must remain indirect or our brains would explode with the volume of data. It's all there in unconscious zip files. You have to be connected daily to intuitive or extra-cognitive ways of thinking and being if you want to utilize this knowledge. Ancestor-mind is worth considering as a kind of Paleo thought-path to take you where you need to go and help you tap into this unconscious knowledge system. If you think about the times when you have attained this altered state in your life, through a near-death experience or peak-performance moment or meditation or just through doodling thoughtlessly on a page, you may recall that complex decisions and activities in the hours afterward were effortless and elegant. You were performing feats of rapid calculation that would crash a quantum computer, accessing a lifetime of memory all at once.

You did this without conscious thought, because to attempt it consciously would fry your entire neural system like an egg. This incredible ability is a gift from your Paleolithic

Ancestors, who had the time and liberty to live within this heightened state of mind every day.

I have previously talked about civilized cultures losing collective memory and having to struggle for thousands of years to gain full maturity and knowledge again, unless they have assistance. But that assistance does not take the form of somebody passing on cultural content and ecological wisdom. The assistance I'm talking about comes from sharing patterns of knowledge and ways of thinking that will help trigger the ancestral knowledge hidden inside. The assistance people need is not in learning about Aboriginal Knowledge but in remembering their own.

Displaced Apostrophes

I'm a guest speaker addressing students and teachers in a school assembly hall, standing in front of an enormous banner with the lyrics of "Advance Australia Fair" painted on it:

> Our land abounds in natures gifts of beauty
> rich and rare.
> In historys page let every stage advance
> Australia fair.

I point to this and say, "I find this offensive," and the room goes horribly silent. I draw out the discomfort for a few delicious seconds more, before eliciting a collective exhale of relief and ripple of laughter from the audience when I reveal that what actually offends me is the absence of apostrophes in "nature's gifts" and "history's page."

I'm a stickler for grammar and punctuation, and mis-

placed apostrophes drive me nuts. Apostrophes are important because in this language they tell us who things and
people belong to. How do you know who you belong to if you
can't get the language protocols for belonging right? I have
been known to lose my head in fruit markets over labels that
say things like "Grape's." *Ooh, you bastards, I want to see the*
owner, and unless his name is Grape I'm gunna get wild in here!

Back at our motel after a full day of speeches and meetings and workshops about Indigenous literacy, us-two see a
professionally printed sign on a brass plate: STRICKLY NO
RUNNING. This makes us question whether the official narrative of literacy interventions in Aboriginal communities is
correct; perhaps low levels of literacy do not represent the
main barrier to our economic participation and advance-
ment.

Strickly no running. Us-two wonder what the prosperous
illiterates responsible for that sign have that we don't. Maybe
it has something to do with belonging to a family that didn't
have its wages confiscated by the government and never returned, allowing for some kind of intergenerational inheri-
tance of capital. Maybe it's about belonging to networks of
informal merchant guilds for settler entrepreneurs. If the
people responsible for that sign are able to thrive enough in
the marketplace to buy motels and sign-writing businesses
while having such shocking spelling skills, surely economic
success must be dependent on factors other than reading
and writing and counting. Whatever it is, that's the secret we
need to be teaching Indigenous students if we really want to
start closing the gap.

Improved literacy scores aren't going to help you if your

community's role in the marketplace is that of commodity rather than vendor. I'm told I'm a decent enough writer, but I've never been able to afford a new car let alone a motel off the back of this hard-won skill. I'm starting to wonder if I've spent decades working on a discipline that has no market value at all. On top of it all, I'm beginning to suspect that literacy causes brain damage, if Plato's opinions on the matter are to be taken seriously.

Those old Greek philosophers worked mostly with oral language. Even when they started writing everything down, they often presented those texts as dialogues, a record of yarns they'd had. Plato told the story of the invention of writing, which was handed to an Egyptian pharaoh by an entity named Thoth. The pharaoh was horrified and lamented that this invention would be the death of human memory, but Thoth insisted, and the demise of oral tradition began. Many more details about this ancient cognitive coup were probably written down over the millennia, but all of that knowledge was lost with the demise of the Library of Alexandria. The moral of the story is to always back up your data if you are committing memory to modes outside of an oral tradition.

Modern neural science has been able to map the way print literacy rewires the human brain. It is a fairly catastrophic process, rearranging neural networks and connections between different areas of the brain in ways that are inefficient at best and highly abnormal at worst. I'm not trying to discourage people or communities from pursuing literacy skills, by the way. I'm not rejecting an entire way of knowing and cultural tradition out of hand—that would be advocating ethnocide, which is never a good idea. I'm merely suggesting that it's helpful to mix things up a little, avoid putting all your cognitive eggs in one basket, keep your brain functioning more optimally. Further, you should never commit all of your cultural knowledge to a print or digital repository. Archives are great, but they are only temporary. The Egyptians learned that the hard way.

The only sustainable way to store data long term is within relationships—deep connections between generations of people in custodial relation to a sentient landscape, all grounded in a vibrant oral tradition. This doesn't need to replace print, but it can supplement it magnificently—those two systems might back each other up rather than merely co-exist. Relationships between systems are just as important as the relationships within them. Oral traditions grounded in profound relationships represent a way of thinking that backs up your knowledge in biological peer-to-peer networks and provides a firewall against dictators who might decide to burn down your libraries. It also mixes things up cognitively and allows your brain to rewire itself in more healthy ways. I call this way of thinking kinship-mind.

Kinship-mind is a way of improving and preserving memory in relationships with others. If you learn something with or from another person, this knowledge now sits in the relationship between you. You can access the memory of it best if you are together, but if you are separated you can recall the knowledge by picturing the other person or calling out their name. This way of thinking and remembering is not limited to relationships with people.

The kinship-mind symbol shows the connectedness between two things—places or people or knowledge or any combination of these. Maybe even synapses. The two distinct entities form a pair, connected by a relationship represented by the line at the center. Additionally, each entity is connected out to a multitude of other pairs, and so the relationship is dynamic, responsive to shifting contexts. There is a tension and balance maintained between the individuality of each entity in each pair and their interdependence in a network of pairs.

In Aboriginal worldviews, relationships are paramount in knowledge transmission. There can be no exchange or dialogue until the protocols of establishing relationships have taken place. Who are you? Where are you from? Where are you going? What is your true purpose here? Where does the knowledge you carry come from, and who shared it with you? What are the applications and potential impacts of this knowledge on this place? What impacts has it had on other places? What other knowledge is it related to? Who are you to be saying these things?

In our world nothing can be known or even exist unless it is in relation to other things. Critically, those things that are connected are less important than the forces of connection between them. We exist to form these relationships, which make up the energy that holds creation together. When knowledge is patterned within these forces of connection, it is sustainable over deep time.

Kinship-mind is one of five different ways of thinking us-two have examined together so far in our yarns. It might be helpful to summarize these ways and define them.

This is the image for **kinship-mind**, which is about relationships and connectedness. In Aboriginal worldviews, nothing exists outside of a relationship to something else. There are no isolated variables—every element must be considered in relation to the other elements and the context. Areas of knowledge are integrated, not separated. The relationship between the knower and other knowers, places, and senior knowledge-keepers is paramount. It facilitates shared memory and sustainable knowledge systems. An observer does not try to be objective but is integrated within a sentient system that is observing itself.

This is the image for **story-mind**, which is about the role of narrative in memory and knowledge transmission. It is the most powerful tool for memorization, particularly when connected meaningfully to place. This is how songlines have worked in Australia for millennia to store knowledge in stories mapped in the land and reflected in the night sky. It includes yarning as a method of knowledge production and transmission. Today it is also about challenging grand narratives and histories.

This is the image for **dreaming-mind**, which is all about using metaphors to work with knowledge. The circle on the left represents abstract knowledge, and the circle on the right represents tangible knowledge. The lines above and below represent communication between these physical and non-physical worlds, which occurs through metaphors. These are images, dance, song, language, culture, objects, ritual, gestures, and more. Feedback loops between the worlds must be completed with practical action.

This is the image for **ancestor-mind**, which is all about deep engagement, connecting with a timeless state of mind or "alpha wave state," an optimal neural state for learning. We can reach this state through most Aboriginal cultural activities. It is characterized by complete concentration, engagement, and losing track of linear time. Ancestor-mind can involve immersive visualization and extra-cognitive learning such as revealed knowledge in dreams and inherited knowledge in cellular memory.

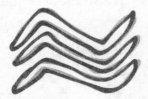

This is the image for **pattern-mind**, which is about seeing entire systems and the trends and patterns within them, and using these to make accurate predictions and find solutions to complex problems. There are three lines with three sections. Each section represents the line from the kinship-mind symbol, which is two elements linked by a relationship. You can see at each point a new pair begins, linked by a new relationship. It is about truly holistic, contextual reasoning.

◉

Pattern-mind links back to the beginning, to the first symbol of kinship-mind, to the assertion that everything is interconnected. Mastery of Indigenous epistemology (ways of knowing) demands being able to see beyond the object of study, to seek a viewpoint incorporating complex contextual information and group consensus about what is real. This is the difference between oral and print-based cultures.

Oral cultures are known as high-context or field-dependent-reasoning cultures. They have no isolated variables: all thinking is dependent on the field or context.

Print-based cultures, by contrast, are low-context or field-independent-reasoning cultures. This is because they remain independent of the field or context, focusing on ideas and objects in isolation.

Plato may have had something to do with the loss of contextual reasoning in Western civilization. He introduced the idea of studying each idea as an entity in and of itself, disconnected from the rest of the system. This practice spawned the scientific method of reductionism and the highly individualized ways of thinking that came to replace more communal approaches to knowledge in Western philosophy. Plato mentored Aristotle, who in turn mentored Alexander the Great, who in turn rampaged east with a great army, installing these new ways of thinking in the high-context cultures he subjugated along the way. Later, the spread of print literacy throughout the West would allow the individual expression of ideas without dialogue, and even individual words to be examined in isolation, causing reductionism to take off like a bushfire.

The low-context ways of thinking that rippled out from Greece and Macedonia proved useful in creating a more obedient workforce and soldiery. People were able to focus solely on the task at hand, rather than the purpose of their work in the greater scheme of things. They did not need to understand the goals of their leaders anymore. Consensus and consent are unnecessary items in low-context cultures. Reasoning is hierarchical, solitary, and disconnected, making it possible for communication to be one-way in the form of rants, instruction, and, most important, orders.

Conversely, high-context cultures demand dialogue and complex agreements. They use a lot of nonverbal communication and leave many things unspoken due to common shared understandings and established consensus about the way things are done. Low-context cultures rely more on extensive verbal communication and explicit, detailed instruction between individuals with relatively little shared understanding. I think this is a nicer way of saying it than the usual "White people talk too much," which I find a bit reductive and also just plain rude. It is an ironically low-context statement for an Indigenous person to make.

It is also inaccurate, as skin color is not a reliable indicator of high- or low-context cognitive orientation. Scots and Russian communities have been found to exhibit the same kind of high-context reasoning as Aboriginal communities. It has also been found that communities can lose their high-context cognitive function through intensive engagement with the global economic system, a shift in ways of thinking that has been measured by psychologists in studies of communities in China and Mexico, which in recent decades switched dramatically from holistic to analytical reasoning processes. Interestingly, Aboriginal communities have been found to retain their customary patterns of cognition whether living in remote areas or urban environments. Limited inclusion in the economic system to date might account for this. Thank you, racism!

So economic activity can be a factor, while skin tone is certainly not, but the main contributing factor is child-rearing practice. In high-context cultures, babies have many caregivers in extended families, while in low-context cultures there are only one or two primary caregivers. The high-context babies eat, sleep, and go to the toilet according to biological demands from moment to moment, while the others are often subjected to clock-based schedules for these activities. They are also restricted by inhibiting clothing and quiet, controlled environments with low sensory motor stimulation and limited adult social interactions, as opposed to the socially dense and unrestricted environments of high-context communities.

Children raised in low-context cultures are exposed from day one to reasoning based on conceptual cues and structures determined by an unknown authority controlling time and space in their reality. This is later reinforced by print-based reasoning in institutions from the age of four or five onward.

High-context thinkers encountering schooling for the first time are fish out of water. They arrive with adaptive, collective, complex, and intensely situated logic patterns. They operate within constantly negotiated feedback loops in reciprocal dialogue with the people and environment around them. They are intensely curious and highly active—traits that disrupt orderly classrooms—and are punished swiftly for it. They tend to sequence things from east to west in alignment with solar movement, as distinct from the left-to-right sequencing of their classmates from print-rich environments. It takes time and patience to drill these orientations out of them, and the process often fails because the kids tend to resist it.

But those who comply can indeed achieve equally in their literacy scores. I have a niece who came top of her class for literacy and is regarded as an educational success story. She can read any text you give her beautifully and fluently, and it makes you proud to hear it. It's heartwarming and makes good content for promotional videos hawking back-to-basics approaches to Indigenous education. She has no idea what any of the words she is reading mean, though. She can't interpret what the texts are about unless there are pictures to help her guess. She doesn't know that she lives on a planet, has never wondered where electricity comes from, and has never heard of Captain Cook.

She is a model student who will no doubt one day become a model worker and a go-to consultant in the design of land-use agreements and interventions. She is none of those things to me. She has a big heart and loves Jesus and wants to heal people, and she is the best person I know. She has, however, been shortchanged in her education. She may have an uncle who is happy to tell her stories about electricity, the solar system, and Captain Cook, but that's not nearly enough to even the playing field for her.

For this chapter I made two beefwood boomerangs in the shape of apostrophes. They both come from the same branch, which was split down the middle to make twinned objects connected by the life force of that tree no matter how far apart they are. I clap them together to make a rhythm, the rhythm of the ax striking the wood as I chipped away and shaped them. There are little stories about ibis in them to remind me of the Egyptian god Thoth and to link that to everything I've learned about the death of memory. I hold them apart and see the spirit reverberating between them. I let my eyes go out of focus and ignore the boomerangs now.

I follow that spirit of connection out to the place I cut the branch back home up north and the relationships I have there. I see the termite mounds, oversized now because the place is out of balance and sick, although to tourists taking photos it looks like untouched wilderness. The parrots that used to lay their eggs in the mounds are gone now, because the moths that used to lay their eggs in the same nest are gone as well. They used to hatch at the same time, and the larvae would eat the waste of the baby parrots. After the moths were wiped out by cane toads, all the newly hatched parrots drowned in their own shit.

There are a thousand interconnected threads like this in the oral culture text I made in the creation of those boomerangs, and I can access them any time I want to pick them up, sit for a bit, and start downloading. I can only describe a few of those threads in this printed text, and it can never be more than a limited translation. I also can't continue modifying and growing this print text after I have written it, or negotiate its meaning through feedback loops with other people or termite mounds or beefwood trees.

Carving those boomerangs is an act of composition that preserves my oral culture patterns of reason and ensures the knowledge that is shared with me has longevity over deep time. Writing this book has the opposite effect, but it is still worth doing. I need to do both to negotiate a viable way of existing in this era of transition, then pass the knowledge on so that others can use it, adapt it, grow it.

It has been difficult trying to carve in the city since I moved here three years ago, though. People get nervous when they see a brown guy with a beard and an ax. On one of the occasions when someone called the cops, I had to hide out for twenty-four hours while a full terrorist response unfolded. On another occasion I was reported to campus security with a complaint of reverse racism, on the grounds that non-Aboriginal people would not be able to participate in the same cultural activity. You are probably imagining some stereotypical old-school Anglo racist in this scenario, but it was actually an Aboriginal person who made the complaint. But enough about silly people—us-two should yarn with somebody remarkable rather than letting narcissists suck away all our thoughts.

I yarn over a few months with Melissa Kirby, a Wayilwan Ngiyambaa woman from Brewarrina who is the deputy principal of a remote Aboriginal community school. She is researching homegrown approaches to literacy, designed by Indigenous teachers, that incorporate oral culture ways of knowing. She also works on preserving and teaching her traditional language. Her mother has developed a phonics system for early literacy that matches all of the sounds in the English language with Indigenous kinship roles, making early reading and writing an exercise based on relationship patterns already known to her students.

I suspect Sisi Mel's been reading a bit of the French sociologist Pierre Bourdieu because she talks tentatively about literacy in terms of transformation of capital. She says the process is about using literacy to transform cultural capital into objectified capital (in the form of assessments and tests), which are then transformed into institutional capital (in the form of certifications) that has exchange value in the economy. The problem is that in our Aboriginal communities we have learned that certifications for us seldom have the same level of exchange value that they have for everybody else, particularly in our home communities. As a result, we struggle to transform that institutional capital into financial capital. We struggle to find the motivation to delay gratification for twelve years to achieve certification, when most of the time it proves to be worthless to us. Literacy programs don't do much for us unless they are linked to radical social and economic transformation in our communities.

I know Melissa is fairly conservative in her political views, so it makes me laugh to hear her talking like a Marxist. The problem is that only socialists have written about the complex interplay between literacy, social fragmentation, and economic marginalization that she needs to address in her community if she wants to improve outcomes for her students. As a result she sounds like a cross between Noel Pearson and Che Guevara.

In our yarns she introduces me to a Cuban literacy program that was tested in the remote New South Wales Aboriginal communities of Wilcannia, Bourke, and Brewarrina. It seems to have been tantamount to an attack on Australian soil in the eyes of the government, which, it is safe to say, will not be taking up the decolonizing model anytime soon, no matter how many gaps it closes. The approach is called a literacy campaign rather than a program or intervention, because it goes beyond the curriculum to build a literate culture and economy in the community itself. It is not about individuals acquiring skills but groups of people empowering themselves, enhancing the cultures and economies of literacy in the community. The Cuban model is grounded in an

Indigenous standpoint of communal rather than individual ways of being, with literacy inequality being addressed as a social problem, not an individual problem.

Sisi Mel likes the approach, but she likes country music and regional netball too much to start wearing a beret and making Molotov cocktails. She is also a single mother with a pretty damn sweet job. She's thinking that after this research project, she'll be finished with all this business forever. In that case, her best bet would be to stop talking to me about Cuba and capital and kinship approaches to literacy. She should relax and listen to Charley Pride on the radio and have a laugh and a yarn. And I can go shooting roos with her brothers on the weekends again like we used to when our lives were simpler, cook up those roo tails, and let somebody else worry about what's working and what's not.

But, dammit, there are two apostrophes in that last sentence, and they catch my eye. *Clack-clack, clack-clack* those beefwood boomerangs say, and it's not my lookout what Mel does, but I might keep digging around here for a bit. Mel's totem is echidna though, and we all know how smart echidnas are and how they like to dig.

Us-two need to go deeper. We might try to find out what happens when we bring together kinship-mind, story-mind, dreaming-mind, ancestor-mind, and pattern-mind in a holistic system of thought.

Lemonade for Headaches

Us-two, we're worried about our little niece. She obviously has acute appendicitis, and she's screaming in pain. We carry her up the road to the clinic, but they send us home with some acetaminophen and dehydration salts, along with a stern admonition to be more hygienic in our kitchen if we want to prevent our children from getting a "sore tummy." We take her back three times before they start to take it seriously, and in the end she is operated on minutes before her burst appendix would have killed her.

Aboriginal stomach pain is often immediately misdiagnosed as a gastrointestinal bug, even though this is one of the few areas of health where the gap goes the other way—statistically, you are far more likely to have this complaint if you are non-Indigenous. This is because the gut flora of Indigenous communities, which is passed to the child from the mother, consists of the microbiome of people accustomed to eating a variety of wild meats. It has been suggested that this is part of the reason why Indigenous populations have until very recently been free of the pandemics of autism, asthma, allergies, and most autoimmune diseases. Cancer and heart disease are also noticeably absent from communities still living traditional lifestyles. Civilized diets have not exactly improved the health of any human populations in recent years.

Not long after our niece's appendix bursts, we go to visit our aunty, who is crippled up with arthritis in her knees and out of her mind with the pain meds she has been sent home from the hospital with. She sends us out to get some leaves of a soap tree, which we pulp in a bucket of water and rub into her knees. In a few days she's up and about, good as new.

While we search for soapy leaf trees we see an old lady at the river, fishing for salmon. The salmon fat in that season is keeping her alive because her kidneys have failed and she can't have dialysis due to recent heart bypass surgery. She cooks the salmon right there on the river bank and eats them all day. She's been told she only has a few days to live, but the fish fat keeps her going for months more until the salmon run passes. The research into fish oil has delivered mixed results, though, so doctors aren't about to start prescribing hand-lines and casting nets for kidney disease anytime soon.

While us-two walk along that river, I tell you about another time when that same glorious old lady was given another death sentence, told she was bleeding out from a ruptured stomach ulcer that couldn't be healed. But we treated that by rubbing my armpit sweat on her face, hair, belly, and arms. In our way, it's the Brolga in my belly that does the healing there, following my scent and slowing the blood so the hemorrhage can stop. The doctors dismissed it as superstitious nonsense and didn't explore the variables further when, ten minutes later, her bleeding stopped and she recovered.

There are discoveries to be made in that ancient practice through scientific inquiry—researchers have already found that the pheromones in male armpit sweat profoundly affect the vascular systems of females, and I'm sure this would have many applications beyond the development of the phero- mone sprays sold in condom vending machines in airport bathrooms that are supposed to trick women into having sex against their better judgment. So far, however, most of the research funding has gone into finding sexual consent loop- holes rather than treating vascular issues.

It takes a dogged commitment to reductionism to ignore so many interrelated variables, to make assertions such as, "Aboriginal people have unhygienic living conditions, there- fore these appendicitis symptoms are probably food poison- ing, so go home and clean up!" Not that holistic medicine practitioners would have been much more effective in any of the above scenarios. Whether you presented with appendici- tis, ruptured ulcers, arthritis, or kidney disease, they would probably all say the same thing: "No wheat, no dairy, no meat, no sugar. Eat plenty of yogurt and couscous [i.e., dairy and wheat]. Take St. John's wort. Two hundred dollars please." The term "holistic" has become something of a dirty word for me, as I've seen it co-opted by charlatans in both health and education. Like "sustainability," the term "holism" has lost its meaning.

So we're left with the devil we know, the narrow focus of modern allopathic medicine—which, apparently, is the reason we are all living longer than anyone in human history. Of course you can't check Neanderthal health records to verify that, or ask a Sentinel Islander without getting shot full of arrows.

Before I start extolling the comparative excellence of Indigenous holistic knowledge and running around like a mad emu on the moral high ground, I should acknowledge our limitations and caution you against assuming that all of our logic automatically conforms to a holistic ideal. First Peoples in Australia experience considerable disruptions and threats to our cultures and ways of life daily, resulting in a lot of dysfunctional reasoning that is not rigorous by any standard. But I usually highlight our best thinking and knowledge, while being careful not to shame people when they stray from this. And we all stray. Silly thinking is something everybody, myself in particular, is guilty of from time to time. It is forgivable as long as you're still listening.

For example, there is a common misconception in many Aboriginal communities that lemonade cures headaches. Every time I hear this I want to say, "Bruz! You're addicted to sugar, and you're crashing right now! That's why you got that headache, and that's why the lemonade fixes it!" But I never do say that. Rather, I redirect that person to some solid item of logic I've heard them use before, such as *Tomatoes are shaped like hearts, and doctors say they make your heart strong. Same idea for kidney beans. I've got a new girlfriend, so maybe I should eat some bananas.* Then I leave them to connect the dots themselves. It is never productive to criticize the thinking of people dealing with the stresses of surviving on the margins of society. We hear about our shortcomings in the media every day—we don't need to hear that from one another as well. But I've included the "lemonade for headaches" example here to avoid offending more economically privileged readers who might like to see a bit more "balance" in this yarn. More important, holistic thinking demands examining things from many points of view, especially those that are opposed to your own.

"Holistic" is a term that is as overused and misunderstood as "sustainability." As there is a relationship between these two ideas, you can't really liberate one unless you reframe the other, so let's put together some holistic ways of thinking and compare them to ways that claim to be holistic and are not, so you can tell the difference.

Let's start out by looking at the recent restaurant phenomenon of Indigenous cuisine. It's marketed as ancient, holistic, and exotic, but in reality it's just a shallow collage of token souvenirs, far removed from anything resembling an Indigenous diet. Customers are titillated by wallaby scaloppine with mango coulis, when it's neither wallaby season nor mango season, and they sit staring at a dot painting over their partner's $50 plate of warrigal greens and magpie goose carpaccio.

This is not Mumma's stingray recipe or *kap marri* dugong. There may be a tiny amount of factory-farmed native meat on your plate, but the culture of the Indigenous-themed kitchen that prepared this frippery has more to do with Parisian traditions than with the customs of the Aboriginal people who work there. Actually, that's not true. In Paris they work with foods that are in season. I don't know what this defrosted *kai-kai* is on my plate.

Emu-fillet medallions with a quandong glaze is not Aboriginal cooking. Aboriginal cooking is not about using native ingredients—it is about using what is available and optimally nutritious at different times of the year and employing cooking techniques that produce the same effect as cooking on hot coals or slowly in the ground. So chicken wings, curry powder, and winter sweet potatoes in a pressure cooker could be considered Aboriginal cooking—kangaroo lasagna is definitely not. Sweet potatoes are only good in the cold seasons after the vines die off, and chickens are too busy mating and brooding in warmer periods to produce good meat. And everybody knows winter chicken is good medicine in the flu season. That's real Indigenous culinary logic—it's about using a holistic method grounded in regional and seasonal knowledge, not just adding "authentic" ingredients. Native foods are out of our price range anyway these days, so we make do with what is available and leave things like *baupal* (macadamia nuts) for more prosperous consumers.

Johnnycakes are another example. Traditionally they are small dampers made from wattle (acacia) seeds cooked over hot coals. Today they are made with wheat flour and baking soda, often baked on a wire grill over an electric stovetop. That is Aboriginal food, adapted to a changing environment and supply chain. Wattle-seed ice cream is not. Even tofu cooked in a ground oven is Aboriginal cooking (although any blackfella caught eating tofu could lose his race card on the spot). A crocodile burger is not.

Aboriginal cooking today continues to reflect the core culinary principles (if not the original ingredients) of our old people—using what is available; balancing animal and plant food, including plenty of long-chain fatty acids; seeking foods in the right season; and cooking them slowly as we would in a ground oven or bed of ashes. I just don't see anything like that in all these award-winning Aboriginal-themed restaurants.

Wild meat is still prized today in Aboriginal communities, but it must be in the right season. Many Aboriginal people will not buy kangaroo meat in stores because it is often harvested outside of the right season and usually has very little fat left on it. This is marketed as "heart smart," but the fat in the right season is very good for you. Good fat is an important part of the Aboriginal diet, as the brain is around 60 percent fat and needs a lot of lipids to function properly. Animals fed on grain rather than grass and harvested out of season do not provide healthy fats. Oil extracted from plants does not provide the healthy fats required in an Aboriginal diet.

A traditional Aboriginal diet is very diverse, with plenty of variety available in each season, delivering precisely the nutrients needed at the right time and in the right place. But with the extensive theft and destruction of Aboriginal lands, there is now a scarcity of wild meat and considerably less time available in our working day to hunt it. Early settler reports reveal how thick the waters once were with fish, how full the skies were with flocks of birds and fruit bats, how heavy the trees with fruit and nuts, how wide the fields of grain and yams. But now, even in the most remote locations, only a pale shadow of that abundance remains.

Our food is now mass-produced on poor soil and has diminished nutritional value, and there are fewer choices in

what we can eat, particularly on a limited budget. Millions
are spent on "healthy choices" campaigns (often patroniz-
ingly using Aboriginal English affirmations like "Deadly!"),
encouraging our people to eat better food. But when all you
can afford is canned meat, rice, and frozen vegetables, there
are literally only deadly choices to be made.

I know we were yarning about health and medicine
before that little rant, but food and medicine, animals and
plants—these categories are difficult to keep separate in a
holistic worldview. There's more to "bush medicine" than
what you may have heard along the lines of "This plant was
used by Aborigines to treat toothache" (always in past tense,
for some reason). It's about a way of living and a way of looking at the world, but above all a method of holistic inquiry.

As an example, us-two might look at a native tree—the
silky oak tree—that has been planted all over Australia in
recent times due to its beautiful flowers. But that tree cannot be examined as a specimen on its own for medicinal and
other uses, because it is part of a complex system, like every
other entity in the universe. That silky oak tree has the same
name in Aboriginal languages as the word for eel. Its wood
has the same grain as eel meat, and it flowers in the peak fat
season for eels, signaling to us that it is the right time to eat
them. Eel fat is medicine in that season and can cure a fe-
ver. The role of plants in Indigenous medicine is about much
more than isolating compounds to be extracted for pharmaceutical use. You can glimpse the true knowledge systems of
Indigenous medicine by looking with a less reductive lens at
things like silky oak trees. This requires more than taking
samples.

Us-two might explore another widespread tree species in a little more depth as an example: the iconic wattle tree. I have carved a fighting *boondi* out of black wattle for this yarn, etched with a design mapping an Indigenous medicine plant garden in Melbourne where I walked daily as I was putting together the knowledge for this part of the book. Wattle has a powerful spirit, and I have a lot of respect for him. He rushes in when land is damaged, like blood clotting in a wound, covering exposed earth and saving the microbiota in the soil. He fixes nitrogen in the ground, and the smoke from his leaves is very useful in many ritual practices. His wood is beautiful and makes good weapons. His seeds are highly nutritious and packed with protein.

When the wattle tree flowers, the wild honey is ready to harvest as medicine—a native honey I know as *may at*, which can kill bacteria like streptococcus on contact. It boosts the immune system and gives you boundless energy. River fish are fat at that time, as are the birds who dive for those fish and show us where to find them. The fruit bats are at peak fat in that season too, and that fat is good medicine for respiratory conditions that might be triggered by the wattle flowers. The leaves of the wattle can be burnt for ash to mix with the leaves of the native tobacco that grows nearby, releasing more of the alkaloids and saponins in the tobacco when chewed. Those compounds are more concentrated at that time than in any other season, making for a wonderful nonaddictive stimulant that enhances concentration and alertness. You need to be alert if you want to find the tiny black bees, which make brief streaks of light as they fly high in the trees when the sun is at a certain angle in that season.

All of these elements combine to form part of a holistic system of good living and good thinking and robust health. But so far, dialogue between the colony and the wattle-flower system has been limited to unsustainable extraction activities. When wattle covers damaged soil on farms to regenerate it, the young trees are poisoned as pests, while superphosphate is strewn over the earth in an attempt to coax some more production from the dead soil. The wattle flower itself is emblematic in Australia—it was used at the Sydney Olympic Games (although the stock was supplied by Canada). Pharmaceutical companies have extracted compounds from native tobacco to make motion-sickness medicine, and the hallucinogenic drug community extracts dimethyltryptamine (DMT) from the wattle tree for recreational purposes. The timber of the blackwood, a wattle, is also valued by guitar makers. You would think a complex system like a marketplace would be able to interact more dynamically with complex ecological systems. This kind of dialogue always breaks down, however, when it is mediated by the cult of reductionism.

Eels, silky oaks, wattle flowers, honey, fruit bats, and tobacco all interacting in reciprocal relationships within a dynamic system of life and knowledge—those are lovely examples but not really transferrable to current medical research and practice in the city. Thought experiment: How can we bring these ideas into a dialogue with science in ways that will actually help?

Perhaps the first step would be a subtle shift in the focus of inquiry to include an Indigenous orientation, examining multiple interrelated variables situated in place and time. We might ask where the fish oil comes from that we are using in our kidney trials and in what season it was harvested. That might explain why our results are so variable and might send our research in new directions. We might even begin to map our reasoning processes onto land-based patterns of thinking in medical research. For example, the well-worn metaphor of the immune system as an army with soldiers, generals, and invaders might be replaced with a more dynamic framework of ants on a riverbank to stimulate some new understandings and directions in immunology.

There are many connected species that deal with dead animals in a river system. There are species in the water (innate immunity) and species on the land (adaptive immunity). A kangaroo carcass (pathogen) on the riverbank is cleaned up by land/air species like crows, flies, and ants, while parts of the kangaroo that fall into the river are cleaned up by water species like eels and yabbies, or crayfishes, to prevent pollution (bacterial infection). The ant totem is related to the yabby totem, so they work in tandem as a pair to keep the system healthy.

The small brown yabbies (dendritic and mast cells) are always crawling about looking for meat. They are the first on the scene, and their excitement on finding the meat signals the blue-claw yabbies (neutrophils) to join the feast. The yabbies need help breaking down the larger parts. A signal is given by the flowering silky oak tree, in that season, for the eels to feast and get fat. So the eels (macrophages or eosinophils) arrive on the scene and help clear the rotting meat from the river.

On the riverbank, the highly adaptive ants (antigen-presenting cells) have been busy too. Foraging ants (helper T-cells) send out scent signals that let other ants and species of carrion eaters know there is a dead animal on the bank. The queen ants (regulatory T-cells) moderate adaptive responses within the ant nests, also with scent signals, responding to dangers like rising water levels (compromised immune system) or opportunities like the right air temperature, to begin mating and to form new nests to feed on different dead animals along the river (diverse pathogens) in the dry season when more animals are dying there. Many ants (B-cell antibodies) swarm over the carcasses to clear the rotting meat (cytotoxic T-cells). The ants also mate with flying female ants who lay eggs (memory cells) for a new generation of ants who all carry the same knowledge of how to keep the river system healthy by dealing with future carcasses (the pathogens you are now immune to).

Understanding biological networks appropriately means finding a way to belong personally to that system. For example, being a Brolga boy, I also have the totems of blood, urine, and the mudshells that you find in mangrove swamps. The mudshell meat is shaped like a kidney and cleans the blood, bladder, and kidneys when eaten raw. That totemic system, which includes other elements like lightning, whirlwinds, and waterlilies, is reciprocal, and I'm in it, mediating healthy ways of being in constant feedback loops of knowledge in that system. Holistic thinking like this is a useful discipline that will not turn up in a tarot deck or emerge rhythmically from your bongos or that didgeridoo you bought in Nimbin. It is also inaccessible to people analyzing shellfish meat in test tubes without knowing where that meat came from and in what season. You have to work at holistic reasoning. You have to grow it from a lived cultural framework embedded in the landscape and the patterns of creation you follow there. There are some general principles of this way of thinking, which we will combine here in some sand talk.

In these sand-talk yarns we have so far looked at five ways of thinking. Let's blend them now into one symbol and one way of thinking, which we can use to create dialogue between scientific and Indigenous Knowledge systems.

In this symbol you can see the shapes of five other symbols for story-mind, kinship-mind, dreaming-mind, ancestor-mind, and pattern-mind. They are not capitalized because I don't want them to become buzzwords absorbed into the marketplace. There are no trademarks in this knowledge. It is not specific to any single cultural group; instead, it belongs to everyone. You should come up with your own words for these ways of thinking if you decide to use them. You should alter them to match your own local environment and culture. This is all open-source knowledge, so use it like Linux software to build what you need to build for a sustainable life. If you want to do this you can use the symbol and your hand now to work through a logic sequence that will help you understand holism and enable you to come to Turtle story later on.

Try pressing your little and ring fingers flat into the desk, or ground, or belly, or any surface where you are reading this—maybe on your arm or on the book itself. Imagine those two fingers are making an emu footprint. Now take the next two fingers and do the same again, but imagine it as a kangaroo footprint. Imagine that each pair of fingers represents a different family group and that the two groups are connected through marriage.

Imagine your little finger now as a child. The child has a singular purpose when it is young—to relate. It relates completely to people and land. This puts children at the center of the family and society, the ones who make relationships happen, tying everything together in a kinship system. So this finger represents kinship-mind—a way of thinking and learning that depends on linking knowledge to relationships with people and with places.

Imagine the ring finger now as a mother. That pair, those two together, mother and child, are the pivotal relationship of any stable society. All other relationships radiate out from, and feed into, this central pair. The first knowledge transmission we experience is through this relationship. So imagine that ring finger now as a mother, telling stories to her small child. This represents story-mind. This way of thinking uses narrative as a device to carry and transmit knowledge and memory in oral cultures.

Imagine the middle finger as a man, belonging to the woman there beside him. He represents dreaming-mind. So the little finger is kinship-mind, using relationships to carry and transmit knowledge, and the ring finger is story-mind, using narrative to do the same. But the middle finger, dreaming-mind, uses metaphors—images, songs, dances, words, objects.

Imagine the pointing finger as the man's brother's child. The man is teaching the child using dreaming-mind, by drawing images in the sand. The woman is telling yarns, transmitting knowledge to the other child using story-mind. Her child is taking on knowledge using kinship-mind—through the relationship and Country the two share together. But the man's nephew or niece, the index finger, is working with ancestor-mind. This means tapping into ancestral knowing that is intuitive and inherent, stored in the body and the land and in spirit, accessed through a peak mental state that allows new knowledge to be absorbed at a phenomenal rate. Ancestor-mind can be achieved through cultural activities like carving, painting, weaving, dancing, and any preparations for ritual or ceremony. It makes you completely open to new and old knowledge.

Now look at your thumb. Wriggle it like a serpent and imagine it sliding across the landscape, forming hills, valleys, rivers, ridges. It connects spiritually to the man, woman, and children. This represents

pattern-mind, which is the skill of seeing the whole
and not just the parts, a big-picture understanding of
how things work. If you can see the whole system and
you have a map of it in your head, then you can see
the cause-and-effect relationship between every tiny
detail. It is the most difficult kind of mind to master.

Now touch the tip of your thumb to the tip of each of
your fingers in turn, starting with the smallest finger:
kinship-mind, story-mind, dreaming-mind, ancestor-
mind. Pattern-mind, your thumb, connects with each
of these in turn. Think about what that means for
each—the vast complexities in our kinship systems
and Country; the way our stories form intricate webs
like maps across Country; the images, objects, and
other metaphors we use to communicate across multiple cultural groups; the ancestral practices and phenomenal feats of concentration required to achieve all
of this.

Now do combinations. Touch your thumb together
with the mother and child fingers simultaneously and
think about the big-picture meaning of kinship-mind
and story-mind together as an intellectual practice.

Now touch the man and woman fingers together with the thumb and think about the pattern and purpose of that relationship, of the generative link between dreaming-mind and story-mind, with story as an extended metaphor. Then with the next pair reflect on the link between dreaming-mind and ancestor-mind, the way cultural activities in peak mental states give rise to metaphors that make meaning and transmit that knowledge with integrity and intensity.

On the back of your hand you will find three joints on each finger. These may help you recall the way our kinship systems can go in cycles of three generations, the way time itself runs in these cycles. In each section of each finger, you may use the creases there to help represent and store increasingly deeper layers of knowledge and understanding about each of these five ways of thinking.

Now make a fist, wrapping your thumb across those fingers and squeezing tight as you think about the way all of these kinds of thinking are intertwined. What is it they form when you put them all together? Don't think too directly about it. Just squeeze your hand in that fist and feel a corresponding squeeze in your belly, and let the question sit with you.

Duck Hunting Is Everybody's Business

In my community there is a thing called *thum pup*. This is a set of two fire sticks nestled in a beeswax pouch studded with giddi beads. One stick is male and the other is female, and friction between them is used to make fire. The female stick is called *thum wanch,* and the male stick is called *thum pam*. Woman fire and man fire. These terms are inverted to make the words for wife and husband—*wanch thum* and *pam thum*. Fire woman and fire man. These spouses share a hearth and bring things to each other there. In many of our cultures, the union of man and woman in this way from the right totemic groups is what keeps fire going in the universe, and without it creation will collapse.

This probably resonates with a lot of people who would like to imagine that the fire stick relationship represents some kind of universal, ancient truth that perfectly mirrors their gendered ideals of family life. But man-woman relations and identities in Aboriginal culture have other layers and complexities, today as well as in the past. There is friction between these sticks, as well as a lot of other kinds of sticks with many different purposes, all of which are vital for sustainable cultures. Some sticks are for fire, others for music or carrying messages. Most sticks are crafted primarily for killing and fighting. Men, women, and children customarily all have access to these weapons because in our culture we avoid the unsustainable practice of concentrating violence into the hands of one privileged group, or outsourcing violence to other places so we can enjoy the fruits of it without having to see it. Violence is part of creation, and it is distributed evenly among all agents in sustainable systems to minimize the damage it can do. We follow creation, so we must all have high levels of competence when it comes to conflict.

And here's where things get messy, because us-two are going to yarn now about the way violence defines expressions of gender in diverse cultures. The traditional Aboriginal relationship with violence is very different from what is accepted as civilized behavior today, and this means that in Aboriginal society there is a different relationship between men, women, and others who are neither or both. I'm not claiming that this relationship remains pure and unadulterated; Men's Business and Women's Business in the contemporary Aboriginal world are contested issues, as we find ourselves increasingly domesticated to varying degrees across our communities. I have tried to write about this before and failed miserably, hampered by the violence of anger and denial that characterizes most contemporary dialogue about gender. While this violence is unavoidable, I didn't want it to be uneven in this book, so I chose to yarn with a black woman who I'm certain could kick the shit out of me if she decided to do so.

There were plenty to choose from. But the woman who stepped into this yarn with me can also smash me intellectually, which she asserts is a far more devastating act of aggression than any physical beating. Kelly Menzel is an Aboriginal woman from the Adelaide Hills and a keeper of ancestral Indigenous Knowledge. She is a nurse by trade and a healer by vocation who is currently completing her PhD and working as a university lecturer.

For this yarn I carved a coolamon, a bark dish that is used to carry both babies and food. I made this one for my woman to carry our new baby son. As with bark shields, coolamons are cut in the right season, then shaped and hardened on the fire before being ground smooth. It requires some violence to be done to the tree, but you cake the raw wound on the trunk with mud, like a salve to help the tree heal. Coolamons can traditionally be made by both women and men, but men will often cut the bark to bring back for women to work on, and ultimately they are very much a woman's object. We can share the process of creating it together, but she owns it.

That's the approach I have taken to writing up this yarn. Kelly decided what was included and what needed to be left unsaid as we dug around in that coolamon and selected what to translate into print. Kelly also filled it up with ideas of her own. She found some massive flaws, places where I had cut against the grain. For example, I asserted that women are physically formidable by nature, but that domestication in civilized societies has weakened them. She didn't disagree with that conclusion—only that I had arrived at it by considering a very narrow range of variables based on unexamined assumptions. She growled at me for my lazy thinking:

> You equate weakness with physical weakness. It is far more complex than that for a woman. Does she also have a sharp tongue and wit and a keen mind? Is she aware of her surroundings? Is she respected only for her physical ability, or is it much more than that? Female power and emancipation are more complex. To be safe, do we need to exert physical dominance over another? Prior to invasion and colonization, women and men were independent and interdependent. Through Law, men and women knew the consequences of both good and bad behavior. The main emphasis was on cooperation and give-and-take between men and women. You need to give some background to these concepts by discussing the difference between Men's and Women's Business.

So we begin our yarn with an explanation of gendered cultural activity in Aboriginal societies. While most daily lived culture and even most ceremonies are shared experiences for all the community, there are also many things that are kept separate. Most fiber craft is done by women, and most stonework and woodwork is done by men. In many places there are activities that only one gender is allowed to do—for example, in some parts of Tasmania swimming and diving have always been the domain of women. There are also many gendered ceremonies that the opposite sex is not even allowed to see, on pain of death.

There is a reason for this division.

Male energy in Aboriginal cultures is different from female energy. Male energy is often described as "sour" and can make a woman belly-sick in many situations. (It is also why crocodiles would rather eat women than men!) In many places a man is required to prepare food while twisted sideways if a woman is going to eat it, so that his belly-power doesn't taint the food and make her sick. (And as this cultural snippet shows, men have always been involved in mundane activities like food preparation, not just the exciting tasks of hunting and fighting.) In Men's Business utilizing belly-power, such as men's ceremonial practice or any activity requiring friction between the hands, women must be protected from that energy. At the same time, women's belly-power can make a man sick and must be avoided at all costs. There are particular circumstances for bringing those two powers together, restricted to acts of creation that keep the generative fires of the universe burning.

But there is also variation and fluidity within and between the genders. In some ceremonies men play a symbolic birthing role, while in others (banned back in mission times) women would strap on massive wooden penises for some dances. Very senior Elderhood comes with a lot of exchanges between Men's and Women's Business, as the men in their old age grow breasts and the women grow beards. Those old people have their own ceremonies that the rest of us are not privy to.

Of course, many people will be born with a gendered nature that leans toward the opposite sex from their biological one, and many others will live on a continuum between one end and the other that may change at different times during their lives. Some people are born with both male and female power in their bellies, allowing them to do useful things that others are not able to do without getting sick.

Not a great deal of lore about this survived after horrified missionaries and early settlers tossed it all out, along with the ceremonial dildos and sky-burial platforms. But a few tantalizing glimpses remain of the particular social roles, for example, of homosexual males who were considered to have special powers in calming distraught children and in mediating disputes. Transgender people have been known in Tiwi society to carry important cultural knowledge and to conduct their own particular ceremonies that are not Men's Business or Women's Business but something else altogether.

Once I was playing around at duck hunting in a public park with a Gamilaraay friend called Luke, a gay guy with a ferocious passion for his culture. I stunned one duck, and a settler woman called out to us, "Hey! Don't you know ducks mate for life? If you kill one, the other will mourn it forever!" At which Luke whipped around and snarled, "Don't project your heteronormative colonial ideas onto my culture!" and I laughed so hard I fell over. The woman laughed too—and the duck got away, so everybody was happy except Luke.

A senior Aboriginal man once said to me about Luke, "It's a pity about him. He's a good fella, but it's a shame about the way he is. In the old days he would have been strangled at birth." This made no sense to me—apart from anything else, how can you determine the sexuality of a newborn baby?— and I lost respect for that man, someone who twists things about the old culture to suit his colonized prejudice.

But we all do that, selectively highlighting some things and playing down others in order to make them align with our worldview. I know I do it. It's pretty much all I've done throughout this yarn, trying to make sense of the most important relationships in my life—which I have to live at the edges of a society that makes sex and gender perilous and complicated. It is less complicated in moments of complete immersion in Aboriginal Law. An old man gives you Goanna story, which asserts the ancient and inalienable Law that rape is a capital offense. In another story, you learn that victorious Aboriginal warriors were once responsible for the first aid and healing of their vanquished foes. Your mum dies, so you and your dad can't cut your hair or beards for a year until the female in-laws give you permission to do so. Or maybe you just make some *thum pup* and listen as the old fellas explain the caring role of a husband. It may indeed be biased to pay more attention to the Law than the recent spate of lawbreakers, but we're looking for long-term trends and patterns in this book, not gloating over short-term pathologies emerging from colonial abuse.

Our kinship systems are based on pairs—uncle-nephew, grandmother-granddaughter, and so on. Knowledge is kept and passed along within these pairs, which have totems relating to them and connecting to land and particular places. A mother might have sparrow hawk totem, which means she also has bushfire totem as the hawk carries burning sticks to spread the fire in that cold season. There are Story Places connected to these relationships and knowledge, and she passes it all along the right kinship pairs to the appropriate relatives.

The dot at the center of this symbol is the child, and the circle around it is the mother. This kinship pair is the center of our society. That circular symbol is one of the oldest in human history and can be found etched and painted on rock all around the world. The purpose of any sustainable society is to protect and nurture this most important of relationships. That is the main role of men as brothers, uncles, fathers, cousins, and grandfathers.

The triangle is the man, who supports the women in his life. But he cannot abuse that trust. Around every woman there are three generations of females: the man's in-laws, who protect the woman and who have power of life and death over her husband. These are her sisters/cousins, mothers/aunties, and grandmothers, represented by the three "c" shapes enclosing the sides of the triangle. In some places a man in mourning for his wife or mother cannot cut his hair until those women send word that it is time to do so, or he will be punished. He is not even allowed to speak to his mother-in-law directly but must practice respectful avoidance. Those three generations form the sustainable structure of kinship, in which there are checks and balances to ensure equality and prevent abuse.

It is true that some abusive practices in Aboriginal society were reported by early settlers, but these early observations followed the cataclysmic spread of European diseases across the continent, decimating our population and wiping out most of the elderly. So most first-contact encounters were with fragmented societies whose governance structures had been severely disrupted.

It is also true that in many cases today our relational dynamics have become corrupted, and some people use the resulting dysfunction to claim that traditional culture is primitive and abusive. But it has not always been this way, and it is not this way for everyone even today. Sometimes people cite examples of community dysfunction to me as proof that Aboriginal intellectual traditions cannot possibly exist. I don't see any connection between these two ideas, except for a desire to diminish (through massive leaps of logic) a culture that is inconvenient to the agendas of development and resource extraction on Aboriginal land. The horrific violence of the occupying civilization is ignored, while our dysfunctional responses to its excesses are condemned.

There seems to be consensus among many Australians that Aboriginal men are abusive, the culture is misogynistic, and traditional Men's Business is some kind of rape cult that needs to be destroyed. Of course there are incidences of abuse—there are in every corner of today's society; it's just that each Aboriginal transgression is recounted in lurid detail in the media. As Kelly tells me, the statistics are undeniable and horrendous, but you need to avoid viewing all of this through a lens of simplicity colored by sensational media reporting.

Kelly explains that Indigenous women are five times more likely to be victims of homicide than non-Indigenous women and that they are thirty-five times more likely to be hospitalized due to family violence. This is higher than the rate for Indigenous men, who are just over twenty times more likely to be hospitalized due to family violence than their non-Indigenous counterparts. (However, we are also less likely than women to seek treatment for injuries.) On the face of it, a person could perceive in the data some fairly brutish patterns of violence in Aboriginal culture, but it is more nuanced than that. For a start, Kelly says, you would need to recognize that over 50 percent of Aboriginal women today have non-Aboriginal partners before making racialized generalizations based on domestic violence statistics.

"Our domestic violence stats are higher not because we are inherently or culturally more violent. It is because we are forced to live within a system that perpetuates violence, creating a sense of intergenerational hopelessness." She adds that there are different kinds of violence, both positive and negative. She says there is a big difference in our culture between controlled, public violence and uncontrolled, private violence. The latter is a very recent development in our communities and is the reason for the horrific domestic abuse statistics.

In our culture not very much of your life is supposed to remain private. Walking off on your own is considered highly suspicious behavior. Violent actions carried out in secret are prohibited, something only done by evil people and sorcerers. Our governance systems are distributed and policed collectively. If there is a dispute, everybody is involved, and if violence is used then it is highly ritualized and witnessed by all. Here's an example.

Us-two are watching a street fight in a remote community. Two Aboriginal women are squaring off like boxers, trading blows with expert precision and devastating accuracy. To me, with my narrow-minded disdain of Western femininity, the fighters are glorious. They do not "hit like a girl" or "throw like a girl" (an action once described to me by a male settler as "how we men throw with our left"). These women have not been raised in confinement, have not had their movements restricted by shame or cloistered activities or clothes that button at the back like an infant's, that might burst open or flap up at the first hint of natural movement. They haven't been taught to occupy less space than males. They haven't been shuttered away from the outdoor adventures and mock battles and free movement of childhood. These women fight like champions. They're half my weight and would drop me in a second.

Six feet away, the partner of one of the women stands with a baby carriage, rocking a baby. He is unconcerned and respectfully averting his eyes from the fight. There is no thought that the baby may be harmed, because no collateral damage is allowed in this carefully controlled violence. The combatants' feet are planted solidly as they rocket jabs back and forth. There is no blood, no serious injury. A crowd, including children, is gathered for entertainment but also to adjudicate.

There will be no hair-pulling in this transparent and public display of justice. There will be no kicking or rolling on the ground. If there is, everyone will know, and the transgressor will be shamed. In most fights there is not even a clear winner or loser, because that is not the point. Everybody has agency in our traditional justice system, and these street fights are its latest permutation in an ongoing struggle to keep our participatory models of governance alive. It is a pale shadow of what once was, and it often fails to produce harmony, but the pattern of old Law is still there, at least in part. I have to acknowledge that when it fails today, it fails with horrific consequences. But the fact remains that in our culture we have a different relationship with violence that is not exclusive to males, and therefore we have a different relationship between the sexes.

In the end the dispute on the street is resolved, and both fighters walk away with their heads held high. The mother of the small child collects the infant from her partner and feeds her. The crowd moves away, but they'll be back when there is another dispute: everybody knows that it is always best to keep violence in the public eye. Behind closed doors, it can turn nasty because there's nobody around to help or bear witness. An extra layer of transparency is added when the fight is uploaded to YouTube thirty seconds later. It joins hundreds of other Indigenous fight videos in cyberspace, most reflecting the reality that in remote Indigenous communities all over Australia, women are fighting more than men are, and they're really good at it. Whether this is a positive or negative phenomenon, it certainly contradicts the dominant narratives about Aboriginal violence. So what is the true story here?

Kelly says there is such a mix of traditional and colonial expressions of violence now that it is hard to identify which is which:

> There is controlled violence versus uncontrolled violence, hidden violence and public violence, violence to resolve disputes, violence born from colonization and dispossession and cyclical violence. There is also an argument to be made for white systems perpetuating nontraditional violence so that members of marginalized groups remain concerned about these things, and not with the decisions that are being "made for us" in a wider sociopolitical sense. I would argue that powerful, conservative white men are afraid of traditional Law so they perpetuate falsehoods about Aboriginal men, women, culture, and knowledges to maintain the status quo.

At the height of the Australian government's recent military intervention into remote Aboriginal communities in the Northern Territory, a powerful meme emerged from the ubiquitous propaganda that was branding traditional Men's Business as a hotbed of sexual assault and pedophilia: "Rape is everybody's business!" This quickly became the motto of every intervention program. Literacy is everybody's business. Employment is everybody's business. Welfare reform is everybody's business. Got a scrapbooking workshop you want to run in an Aboriginal community? Everybody's business! There seemed to be a push to alter Aboriginal culture into a simplified, one-size-fits-all expression of identity, free from the troubling element of Men's Law. Some high-profile

Aboriginal women even began writing about the dangers of Men's Business, calling for an end to traditional culture in the name of protecting women and children, alongside calls to open up traditional lands for development and outside investment.

My female colleagues (both Aboriginal and non-Aboriginal) were handing me these op-ed pieces on a regular basis, as you might pass a rehab brochure to a junkie—with compassion and care, but also a twist of mild disgust and fear.

Fear of Aboriginal men unified Australian politics at all levels—there was even rare bilateral support in parliament for sending the military into Aboriginal communities to protect the children by putting in place ninety-nine-year land leases for resource extraction. There was more public debate on whether gay soldiers should be allowed in the army than there was on suspending the Racial Discrimination Act and sending those soldiers in to occupy Aboriginal communities.

Germaine Greer stepped in to defend Aboriginal men, and hers was a lonely voice. She saw us forced to abandon self-determination for assimilation, in much the same way as women were forced to abandon liberation for equality. She wrote an essay about this and was widely condemned for it, even called a racist. She inspired me, though. I was also inspired when I saw a hashtag campaign run by Aboriginal women promoting strong fatherhood role models in our communities. But these were marginal voices lost in the maelstrom.

On a personal level, the atmosphere created by this media storm of horror stories damaged my relationships with both non-Aboriginal Australians and women. I found myself irredeemably classified as a member of both an abusive culture and a predatory gender—a gender inimical to the strong women who made my life possible, who were suddenly supposed to be afraid of me because on average my group was physically larger. Those powerful women were supposed to relinquish all agency when it came to violence, deferring to me by virtue of my membership in this class with its exclusive monopoly on physical conflict.

This was one of the more distressing periods of my life, but through that pain I came to realize a few things. On a personal level I had to recognize how much I depend on the approval of women—and the tantrums I tend to throw when I don't receive it. I'm still working on that one. More important, I realized it is not enough to examine individual thoughts, words, and behaviors—we need to look at the hidden systems of control that Kelly says are shaping our lives and our pathological responses to these power structures. We need to stop looking sideways at one another to identify victims and oppressors. We need to start looking upward.

The flawed relationship between civilian men and women is the basic unit of our enslavement in the global economic system that has come to dominate our lives. There is no way men would submit to the labor we undertake without needing to attract and then support our women or, worse, without the promise of being able to dominate them by accumulating capital and forcing them into dependence. Ninety percent of the world's wealth is owned by men. Most of the wages in the world go to men, while women do about two-thirds of all the work, most of which is underpaid or not paid at all.

Sustainability is an impossible dream in this unevenly gendered system, so it is worth revisiting the idea of liberating ourselves from it.

This is not an issue of ideology but of sustainability. Arguably, it even represents an existential threat to civilization itself. Every culture in history that has elected to subjugate women has had difficulty sustaining itself for much more than a thousand years. You can't maintain a culture that is based on retarding the development of half the population, particularly the half that is responsible for creating life. Kelly says this abomination has been achieved by altering the cultural blueprints of modern masculinity and femininity:

> These are constructs that don't fit with our culture. Traditionally, our men are tender, empathetic, and nurturing. Our women are strong, determined decision makers and pack a punch should it be required. There is a unique equity in our culture. Whilst we have male and female energies, these do not equate to masculine and feminine in the way Western societies define those terms.

Everywhere civilization goes, most women are excluded from active participation in violence and then domesticated into a twisted, soft, flouncing version of femininity. In Asia, the Middle East, and Europe, in every civilization, women are forced to adopt a passive role, their bodies confined and weakened until they are at the mercy of the men around them. They cannot walk the streets unaccompanied by males for fear of attack. Kelly says this enforced passivity is not just physical but intellectual as well. She and I reflect on how long this has been going on in the world and when exactly it began.

An Elder from Victoria once told me he thought this all began about twelve thousand years ago, when there was a global shift toward enhancing the importance of the role men play in procreation. Before that, many societies were matriarchies, but suddenly men began worshipping bull horns and erecting obelisks and massive stone cocks, along with towering walls and strongholds. Gilgamesh smashed the world beneath his sandaled feet. Agamemnon's wife surrendered her crown to him, and he started conquering and building an empire that would eventually raze the topless towers of Ilium. Heroes like Achilles raped and pillaged their way across the north. Theseus sought out the last priestess of female power, exiled in the labyrinths of Minoa, and lured her away with promises of undying love, only to leave her jilted and heartbroken on a distant shore.

And that's how romance works. It exploits the Achilles' heel of exceptional women: their desire to think the best of men and stand by their side. Contrary to popular belief, men are not turned off by powerful women. Rather, they long for them, court them, wine and dine them, and ultimately either ruin them or lock them in their towers. It was the violence of romance that conquered women, more than witch pyres and swords and pillaging. Once trapped, the protection rackets run by their captors kept terrorized females dependent and compliant so as not to disturb the precarious and conditional security they were offered. They were then fattened up and put to work on their backs, either as breeders or playthings.

Kelly has some more thoughts on this:

The subjugation of women is perpetuated by multiple means. The myth of romance is political. It is a myth about male-dominated hetero couples, where an incomplete woman is completed by her relationship with her partner. Patriarchy naturalizes this sexual identity, masking the cultural construction of the feminine, thereby continually reproducing women in a subordinate position.

Romance is different in Aboriginal cultures. There is a word for it in many of our languages, often mistranslated in mission times as "adultery." It is love magic, that body of Lore that maps the chemistry and lightning charge of attraction and sex. All are equal agents in this game of love, and as long as you don't play with anybody from the wrong kinship group, anything goes. There is no obligation for men or women to be limited to one partner for life, but marriage is still sacred.

Selective recording of observations by anthropologists and early settlers popularized the idea that old men used their power and influence to marry underage girls and monopolize the young women of the group. This ignores the fact that old women did the same with young men—a mechanism that makes sense in terms of maintaining stable populations with low birth rates. It also served to curtail the excesses of youth and to mentor young people in how to navigate the complexity of sexual relationships. You may think of this as pedophilia, but bear in mind that this was in an era when adolescence lasted months rather than years. All over the world, children became young adults and married much sooner in those days.

Selective historical observation also gave rise to the idea of Aboriginal males controlling subservient women with abuse. An instance of a man beating a woman would be interpreted as unchivalrous and barbaric, with projections of "natural" male domination being applied to the situation without context. The real context was that everyone, male and female, could be punished for transgressions, and this kind of justice was policed by all members of the group rather than a centralized male authority. Yes, women were beaten the same as everyone else when they broke the Law, because they were respected—equal and accountable in the eyes of the Law and not regarded as weak and fragile creatures.

"It is also not just about Law and violence," Kelly reminds me. "Breaking of Law does not always result in an act of violence, and once the punishment is undertaken everyone moves on, unlike Western culture where a person carries their crime and punishment for eternity." She also asks me to speculate on what kind of generalizations a settler might make upon observing an Aboriginal woman beating a man. From experience, I can tell her this is often viewed as an anomaly and dismissed entirely.

When I think of the worst public beating I ever received from a woman, resulting in three busted ribs, a knife through my hand, and half my hair pulled out, I recall that the non-Aboriginal observers of that fight ignored the power of that magnificent woman and focused on my weakness as an individual who had somehow let my sex down. The onlookers, both male and female, were in fact so disgusted with my poor performance that they didn't even bother calling an ambulance, leaving me to crawl my bloody way home. Observation tends to be very selective when it comes to making broad generalizations about gender and ethnicity.

A good example of selective settler observation of Aboriginal culture that has resulted in a lot of confusion is of the "hunting and gathering" division of labor between the sexes. It is widely believed, even in our own community today, that traditionally men were the hunters and women were the gatherers. However, in the old stories we see both genders taking part in both activities. Most of my gathering knowledge has been learned from men. Some of my hunting knowledge and most of my fishing knowledge has been learned from women. In communities where hunting and gathering still take place daily, you may see men gathering plant food and women fishing and hunting animals with their digging sticks.

That's another thing—those women's weapons are not called digging sticks in any of our languages, only in English. They are used for many things but are primarily designed for hunting and fighting. Men use boomerangs to dig, but those are not called digging sticks. Only women's weapons have been linguistically domesticated in this way. Try telling old granny up north she's a gatherer, and she might crack you on the head with that fighting stick (before taking you out to catch a fat goanna).

It is true, however, that Women's Business gave females control of nutrition in our communities. They directed most of the harvesting, production, and distribution of food, giving them considerable authority, in addition to controlling the creation and nurturing of life. The role of men was to bring in the things that women and the rest of the community needed. It has never been the cultural role of Indigenous men to subjugate women, or even protect them. Our women have always been more than capable of protecting themselves, as they have always had access to the skills and tools needed to use violence when necessary.

Creation started with a big bang, not a big hug: violence is part of the pattern. The damage of violence is minimized when it is distributed throughout a system rather than centralized into the hands of a few powerful people and their minions. If you live a life without violence, you are living an illusion: outsourcing your conflict to unseen powers and detonating it in areas beyond your living space. Most of the southern hemisphere is receiving that outsourced violence to supply what you need for the clean, technological, peaceful spaces of your existence. The poor zoned into the ghettos of your city are taking those blows for you, as are the economically marginalized who fill your prisons. The invisible privilege of your technocratic, one-sided peacefulness is an act of violence. Your peace-medallion bling is sparkling with blood diamonds. You carry pillaged metals in your phone from devastated African lands and communities. Your notions of peaceful settlement and development are delusions peppered with bullet holes and spears.

Violence exists, and it must be carefully structured within rituals governed by the patterns of creation and the laws of sustainable cultures derived from those patterns. Violence employed in these highly interdependent and controlled frameworks serves to bring spirit into balance and hold in check the shadow of the I-am-greater-than deception. Every organism in existence does violence and benefits from it in reciprocal relationships.

Duck hunting results in duck death. Yam digging results in the death of equally sentient plants. Abusing your partner results in a spear through the leg. Hunting stingray can make you into dinner for a tiger shark. Domesticated beings are stripped of this reality and become passive recipients of violence—either its benefits or its cruel impacts. They devolve as a result.

What would it mean to reverse this domesticated state? It would take centuries to transition from human domestication and recover our exceptional physical and mental powers as a custodial species. It takes a few generations for pigs to get over it when they escape into the bush. At first they remain the fat, pink, stupid beasts they were selectively bred to become over centuries of captivity. But soon they grow black bristles and long tusks, each generation becoming faster, stronger, smarter until the formidable razorback emerges. I often wonder what men and women would transform into outside of captivity.

At least once a day I think of some old footage I once saw of the first European contact with Andaman Islanders off the coast of India. There is a man and a woman, equally lithe and muscled, walking boldly toward the camera with fierce eyes. They are clearly life partners and bear the same height, strength, and status. Neither stands in front or behind—they are side by side. Their skin glows, and they take my breath away. They make me feel like I am a poodle watching the approach of two dingoes.

"Where you going?" they ask with their bright eyes, shining with superhuman health and intelligence.

"I have no fucking idea," I whisper back.

Fortunately, Kelly does know where she is going, so the last word here will be hers:

I don't know if a reversal is necessarily what is required. Reclamation perhaps? I have to admit I do enjoy some of the perks of modernity and contemporary life. And I am certainly more of a lover than a fighter, although I can hold my own should it be required. But I am interested in deconstructing the current perception and understanding of masculine and feminine and reshaping it into something less rigid, less claustrophobic, and less limiting. This would require people to open their minds and spirit to other ways of being and doing, and respecting and celebrating difference. Men's Business is crucial to this process, but at the moment it is feared because it fundamentally challenges white men and the status quo. The burden of patriarchy and misogyny narrows the lives of every person, irrespective of gender and culture. I strongly believe there needs to be a process of acknowledgment, reflection, and renewal.

This requires more than an act of resistance; I see it as a process of rebirth, redesign, and reconstruction. We must move beyond this policed and regulated form of domestication we are now enduring. The boundaries of public discussion need to be widened with information, compassion, and honesty. Collectively, we need to break free from the bondage of patriarchy, white privilege, and the misogynistic structures that control us.

This isn't just desirable; it is necessary if we want to be sustainable.

Immovable Meets Irresistible

Us-two, we're camping about thirty miles south of Darwin, and Oldman Juma is teasing us because he knows we're scared of frogs. It's not really a phobia, just a cultural behavior instilled by marinating in Murri moments of delicious terror over the years. We giggle and wriggle closer to the fire, happily regressing into childhood for a while as Oldman torments us with toad stories. "See that red eye there in the dark? He'll come for you tonight, hee hee!" We can't always be grappling with the complexities of the universe—sometimes we just need to play. Sometimes grappling with the big questions and issues can be playful too, so we sit and yarn and laugh around that fire, seriously humorous.

Bodies in motion, bodies at rest. Unstoppable forces and immovable objects. Big bangs. Every action has an equal and opposite reaction. Ah, but somewhere in between action and reaction there is an interaction, and that's where all the magic and fun lies. See? I yarn about what I call *ngak lokath*, the brackish water formed in the wet season when the fresh water floods down the rivers into the salt water along the coast, an action that the Yolngu call *Ganma*, a phenomenon of dynamic interaction occurring when opposite forces meet in an act of new creation. It is a principle that guides the interactions of different groups and interests in society, a reflection of the natural processes of self-organizing systems.

Changes and transformations occur this way. Dynamic systems of culture, even languages, evolve over time through these interactions. For example, there is no word for "hello" in my home language, and until recently there wasn't in English either. It developed through maritime interactions of different peoples hailing one another from different vessels at sea. We didn't say "goodbye" in English either, until the words "God be with you" were muttered so many times that they became something else. Handshakes evolved from people checking for hidden daggers, then evolved again as subjugated peoples elaborated on the custom, bumping fists and exchanging a complicated series of different grips. Hello—goodbye—*whichway?*

The greeting "whichway?" came into Cape York and Torres Strait creoles when two different laws and languages collided. The regulating protocol of greeting people by asking them where they were going (i.e., "What is your purpose here?") was about enforcing custodial authority of lands under Indigenous Law. The English phrase "Which way are you going?" was adopted as a translation of this, and the new greeting emerged. Later, in the south, the regulatory question asked of outsiders took on a different meaning through mistranslation and became "welcome," which was never a concept in our cultures before the twentieth century.

This is what happens when worlds collide and then mingle over time. Living cultures and languages evolve and transform. Thought experiment: people still say "Pick up!" when calling a busy person's mobile, even though nobody really uses landlines that require picking up anymore. What might future evolutions of that utterance involve? Maybe one day, long after all phones and electronic devices are gone, people will still call out "Pick up!" when they want to get someone's attention. Maybe this will mutate into the word "pikapa" and become the new "hello." It might take on new levels of meaning and cultural significance, mirroring the sound of a new amphibious species, because this kind of symbiosis is the way dynamic transformations work in self-organizing systems. Forces collide and a myriad of symbiotic relations and synchronistic solutions emerge from this vibrant complexity in which nature and culture are never separate fields.

Oldman Juma says this is what is happening with cane toads. He says they have their own Dreaming here now and denies that they are noxious pests anymore. He says their role in creation is to mop up all the toxic pollution of civilization and store it in their poison sacs. He says this story has changed them on his ancestral lands, mutating them in response to the patterns of land and culture. He says some of them walk like cats now and have red eyes and claws. He points to tiny red pairs of glowing orbs beyond the firelight in the bush all around. "See?" I don't quite believe him—he might still be joking around, but I'm right on board with the physics of his Dreaming when he shows us the answer to the question of what happens when an irresistible force meets an immovable object. Everything! He draws it for us in the sand.

These two symbols show creation events unfolding from impact points at their centers. The woman and child symbol we looked at earlier, the circle with the dot at the center, shows a point of impact creating the child. After that, the circle divides, creating the two semicircles of the symbol on the left here. This is a Turnaround event, an action within creation of separating spirit and physical worlds, earth camp and sky camp, to make our reality. This symbol is Sunrise and Sunset Dreaming, showing the continual interaction between night and day, of worlds coming together and moving apart in cycles of creation, beginning and ending, expanding and contracting, like breathing in and out.

It may also refer to the Seven Sisters story of the Pleiades constellation, where one sister is left behind on the earth and returns to the sky periodically. It is an eternal enactment of a drama between male and female. The symbol on the right shows the male side, a diamond shape from the Orion constellation, split by an impact at the center. He is chasing those sisters. You can see the seven spirit families again in the points of this symbol. These two male and female Dreamings interact with each other in a dance of tension and balance, separation and unity. This action happens both within them and between them. Man-woman pair together making fire in the universe. Stories about man-woman transgressions and struggles warn us about what to avoid, to punish, to grow beyond.

Hybridization is a force that influences the patterns of creation. It happens with plants, animals, cultures, languages, and laws, which are all part of the same nature. When this hybridization is occurring organically, through the interaction of a multitude of agents in a sustainable system, the emergent entities are usually productive. Abominations occur when hybridization is forced unilaterally by an outside agent. A ridgeback crossed with a pit bull is probably not a good idea. Mind you, neither of those selectively bred, former-wolf mutants was a good idea in the first place.

There are a lot of opportunities for sustainable innovations through dialogue between Indigenous and non-Indigenous Knowledge systems that might help civilization transition into sustainable ways of living. The problem with this communication so far has been asymmetry—when power relations are so skewed that most communication is one-way, there is not much opportunity for the brackish waters of hybridity to stew up something exciting. In cross-cultural dialogue, the most mundane ideas can become pivot points of transformation.

Let us-two do a thought experiment in dialogical thinking using a boring administrative example. How about risk management? From an Aboriginal point of view, this might be calculating the risk of being taken by a crocodile when crossing a river. The first time the risk is minimal, but if you're going to do it twice, the risk is greater—and you are not just doubling the risk. The crocodile is not an abstract factor in an algorithm, but a sentient being who observed you the first time and will be waiting for you the second time. Your risk has multiplied exponentially. From a non-Aboriginal point of view, risk is a discipline that is currently based on models that ignore such complexity. It is assumed that risk is random, with probability calculated in ways that are disconnected from chains of cause and effect. Above all, risk is seen as something to mitigate by preemptively providing for compensation before the event. So in the stock market, for example, investments can be insured against loss, then those insurances can be bet upon, with the outcomes of those bets insured as well, on and on into infinity like Russian dolls. In a cross-cultural dialogue, we might see that the problem with this model is that every time you create a new layer of derivatives in this way, you double the size of the system, but you do not merely double the risk—you multiply it exponentially.

Further, subsidizing corporations and providing tax breaks and bailouts and low-interest loans does not minimize the risk of recession and distribute the fruits of this largesse throughout the economy, as is often claimed. The corporations just take that capital and buy back their own stock when it crashes, creating artificial demand for items of diminishing value before selling at the top and starting the cycle all over again. That's the risk equivalent of swimming up to a nesting croc and telling her that her babies are ugly.

In dialogue we might see that the biggest problem with contemporary approaches to risk is the illusion of safety as a human right that can be controlled as a variable in advance. It cannot. In fact, there is no such thing as safety in Aboriginal worldviews. We have no word for it in our languages. Safety provided by an invisible hierarchy is complete anathema to our way of being. There is no agency in safety, which places a person in a passive role, at the mercy of authorities who may or may not intervene when needed. So we have no word for safety or risk. However, we have plenty of words for protection.

Protection has two protocols. The first is to look out for yourself. The second is to look out for the people around you. This is such a wonderful way to live, knowing that you have the power to defend yourself and the ones you love, while also being intensely aware that at any given moment there are dozens of people who are watching your back as you watch theirs. This is the interdependence that our kinship pairs and networks of pairs offer. It is a useful algorithm that could be applied to risk management in the financial system.

The trouble with this thought experiment is that we're doing it on our own, and the kind of dialogue that harnesses the power of hybridity doesn't work that way. There must be a lot of yarners and yarns to make it work properly. It's also coming from an asymmetrical relationship, with me assuming a position of moral authority over other points of view.

That kind of thinking doesn't produce complex solutions and dynamic innovations. But some interesting things can come from uneven power relationships too. In the end nobody can stop dialogue, because it is a force of nature. It emerges eventually, in the form of resistance or disruptive innovation, if you don't allow it expression on an even playing field.

I yarned with Oldman Juma's son-in-law, a bush lawyer from the Northern Territory, to find out about some of the legal mutations that are emerging at the point where the laws of the occupiers and the Law of the occupied intersect. Actually, I couldn't really call it a yarn because you can't get a word in edgewise with bush lawyers. The only thing I contributed to the exchange was the occasional grunt of agreement or confusion. Bush lawyers have been popping up all over the place in the last decade, with many of their ideas seeming to incorporate American alt-right militia "strawman" and "freemen on the land" philosophies.

So much of that theory seems dodgy to me—like the notion of birth being related to the berth of a ship under admiralty law, somehow making your birth certificate simultaneously invalid and a legal contract against which you can claim millions of dollars to pay your bills. Despite that weirdness, the phenomenon is certainly a fascinating disruptive innovation. Oldman Juma's son-in-law claims to have had over a dozen cases dismissed using his particular brand of this innovation, arguing for dismissal on the basis of lack of jurisdiction.

At the most basic level, his legal argument begins with the outcome of the *Mabo* case, which found that Indigenous Law, title, and sovereignty were not extinguished by European occupation. By extension, Indigenous Law is still in place, so the Australian legal system has no jurisdiction here. But that is just the beginning. He goes further and argues that Australia has no rule of law and does not legally exist.

The argument goes something like this: because the Commonwealth of Australia is registered and listed as a foreign-owned corporation by the US Securities and Exchange Commission, court representatives employed by the state can only be third parties acting as service providers for the trustee of a defendant's contract, which is the Australian Constitution.

The defendant can only be tried by peers, who would need to be sovereign people under Aboriginal Law if the defendant asserts that Law. Further, the case is being brought against a fictional character created by an illegal state through the printing of a birth certificate. The defendant argues he/she is in fact a flesh-and-blood sovereign being. As that sovereignty was never ceded and no treaty created upon the illegal occupation of Australia, the Indigenous defendant is living in a community that is under martial law.

The defense asserts contract law, then challenges the judge, asking for the name of their creditor and accusing them of impersonating a Crown agent and presenting an unconscionable contract. It is then asserted that Australia is under the jurisdiction of common law and international law and even the jurisdiction of the US Securities and Exchange Commission, where it is listed as a corporation that is trading insolvent and bankrupt. (There is some logic I don't quite understand here, whereby the defense claims rights and protections under the Constitution but also claims that the Constitution is invalid. But legal arguments always seem circular and contradictory to me.)

The defense then cites a number of provisions of the Crimes Act, asserting that the defendant does not give consent for the matter to be heard in a court of summary jurisdiction. Or something like that; this bush lawyer talks too fast for me to follow. The only documents I've seen from him are in all caps, which I associate with ill-conceived manifestos. But then, a lot of people think of my writing in the same way, so who am I to judge?

Besides, Oldman Juma's son-in-law claims to have won more than a dozen cases this way, with the big reveal at the end when he asks the court to produce documents showing jurisdiction over his Aboriginal client. He informs the court that they can't acquire jurisdiction if they are in breach of administrative law and that if they proceed they are admitting liability. He then asks for the case to be dismissed, and apparently it works.

I have tried to find out from several legal professionals whether this argument has any validity in law, but they won't talk to me about it beyond a red-faced "Absolute nonsense." I ask them to tell me the points of law that contradict the argument, and they just say, "Ridiculous," and walk away. Maybe it's a bit like when people ask me if my Ancestors ate their children—not a conversation worth engaging with. Either that or there's something in there that scares the hell out of them.

What I am interested in here is not some legal loophole, but the fascinating hybridization occurring where the legal system meets an involuntary minority that is having to find creative ways to deal with its inequities and iniquities. There are plenty of both in the Northern Territory, where in late 2018—when I had this yarn—there was not one single non-Aboriginal person in juvenile detention. *All* the inmates were Indigenous. If you've ever met a fourteen-year-old bogan from Darwin, you'll agree that those settler kids are far from being civic paragons, so something is not quite right with the justice system there. No wonder the Indigenous community is producing some disruptive innovation in the courtroom.

Indigenous challenges to Australian law are not, however, limited to fighting the tide of black multitudes being incarcerated for nonpayment of fines and other minor offenses. Some Aboriginal groups are getting on the bandwagon of global trends of secession and are working toward a massive Blexit (like a blackfella Brexit). I yarned about this with Ghillar Michael Anderson, founder of the Indigenous Sovereign Union movement.

Ghillar was named for the hundreds of galahs (cockatoos) that flocked screeching around his house at the moment of his birth. On ancestral lands, Galah is a creation entity who made a new kind of boomerang in partnership with a frill-necked lizard, who then turned on him in jealousy and smashed his head with that boomerang. Galah responded by knocking him into a big patch of bindis—spiky seeded plants. This story always makes me laugh because Oldman Juma refers to narcissists as frilly lizards running madly about and trying to look bigger than they really are. Our old Law is all about containing the excesses of such people, but it's not a job that's getting any easier today because there are millions of them. Along with his

name, Ghillar also seems to have inherited the totemic role
of keeping the Law in place to resist those insane enough
to think themselves greater than the land and its peoples.
The work he does for the Sovereign Union is about uniting
First Peoples under that old Law of the land and asserting
independence.

I share with him my concerns about our mob taking on
Westphalian models of nationhood, and his initial response
does nothing to allay my fears. He talks about the Treaty of
Westphalia coming out of a medieval Europe, of kings and
barons recognizing each other's boundaries and authority to
put a stop to the depredations of warlords. He says we need
to do the same thing in Australia with a treaty recognizing
Indigenous heads of state, kings and queens of Country.

He reads unease in my silence and changes gears to help
me understand. He has spent a lot of time on Cape York, so
he puts it all in the context of people and stories I'm familiar
with. He has spent decades working with resistance movements in places like Kowanyama and Lockhart River, and
he was at ground zero in Aurukun when the Queensland
government took control of the community from Aboriginal
leadership in the late seventies.

In those places, like everywhere, dispossessed Aboriginal
people were brought in from other places and left there.
Now, a lot of them didn't marry or adopt into the local families, so they didn't come under that Law. If the families of
Traditional Owners don't keep that authority for Country,
who will speak for them, look after those living away from
their own country? Who can make sure they don't do the

wrong thing? In most cases those dispossessed ones can't be repatriated because they have nothing to go back to, so you have to set up your own structure defining territory and boundary to make sure people off-country are taken care of. So they can have the right to a say in the community but live under the authority of the Traditional Owners. Then they can be adopted or married in the right way under traditional Law, and we can bring everyone home again.

He refers to the *Mabo* case, in which Eddie Mabo had his title to inherited estates recognized when he was a family member not by blood but by customary adoption Law. I understand this well because of my own induction into the protocols of customary adoption over the years, and I understand the importance of taking on those names and obligations in coming under that strict Law when you have been displaced. Half of our people are living in a kind of purgatory because they haven't been brought into their host communities under a framework of Aboriginal Law in this way. Ghillar then eases my concerns about the potential of sovereign-nation movements to undermine the old ways with new hierarchies:

It won't corrupt the old way. The *Mabo* case recognized that old Law—not the same structure of English law but inalienable by another power. If we start using their hierarchical models, they'll beat us every time. You have to keep the old structure to keep our Law inalienable. Use the foundations of that and adapt it, so it underpins the new way of making decisions in a Continental Common Law, following song-

lines and doing Ceremony because that's the one that holds
it all together. We all talk about the Creation Serpent, but
we give him different names. But he's still the same one. We
may speak different languages, but we still have a common
Law Story and belief.

Ghillar recently spent three days with international
lawyers studying the legal processes of the decolonization
of India and other former colonies, finding precedents and
flaws. He has also been able to establish that the legal code
installed by Governor Arthur Phillip in Australia situated
the colony under the governance of criminal law and the
rules and disciplines of war. He says that no law exists to
take away the rules and disciplines of war, so those principles still apply to the governance of Australia under section
51 of the Constitution. Ghillar explains that this is why we
have a police force rather than a police service, as the police are a military force operating under the guise of law and
order while carrying out a war against Indigenous Australia. He says that, under international law, we need to have
a treaty or compact to bring an end to hostilities before we
can have any legally valid negotiations, and that there is
some urgency to do this:

> Over the next three years there is a very big push for con-
> stitutional recognition to swallow our sovereign rights in
> time for Prince Charles to become king and then declare
> Australia a republic. We need to make our move to identify
> and map our Country and treaty with each other and form a
> union of nations to have a powerful position.

Ghillar has also shown native title to be a massive sleight-of-hand trick to extinguish Aboriginal Law:

> In every case we have to prove connection to our old Law at the first moment of British sovereignty over Australian soil. So we are tricked into handing over our sovereignty at the very moment we declare it, by recognizing theirs. We then give everything away through Indigenous land-use agreements. We're not told what it means. We sign papers that legitimize settler land grants, extinguishing native title.

He has been working with groups in the Kimberley region to wind back native title, transferring powers out of prescribed corporations and into community organizations independent of the Native Title Act. People are skipping the native title process now and negotiating directly with developers, rather than working through lawyers who are supposed to represent them but are secretly being paid by mining companies. He says he is exposing all kinds of shady deals and Elder abuse in this process in the Kimberley, where there is a projected $17 trillion worth of mineral reserves.

The Sovereign Union is seeking unification with other Indigenous nationalist movements, such as the neighboring Murrawarri Republic, which straddles the border between New South Wales and Queensland. I was introduced to that group by Mumma Doris Shillingsworth, who calls me her son whenever I'm on her country so that I am regulated by her Law and bound by appropriate protocols. Their vision is inspiring, but it runs into some PR problems when some of the old people ask, "What will happen to my pension when we kick out the commonwealth?"

Some members have visions of mining the resources on their land to pay for the Murrawarri Republic. They aim to go off the power grid by installing solar panels. They plan to make solar energy infinite and sustainable by mining the rare earth metals needed to build the solar panels on their own land. When I ask them where they plan to store the radioactive waste from refining those rare earth metals, I realize too late how cheeky and disrespectful I'm being. I've even asked how we will resist accepting loans from corporations to build infrastructure for these mining projects, losing our sovereignty all over again like so many decolonized nations around the world. It's not my place to ask these things. I've overstepped, and I feel shame.

I hunt a fat pig and bring it back for them to cook up, and they are slightly mollified, but I'm not invited back again after that. I remain troubled by the potential risk of creating Indigenous civilizations, Anglo economic systems administered by men with black faces but still following the same unsustainable global blueprint of destruction. It might be advisable to take a moment to listen to the warnings of Ghillar and ground our strategies in our old Laws of sustainability before we go charging off and building a Mumbai on Aboriginal land.

The object I carved for this chapter was a very intricately inscribed killer boomerang cut from a messmate tree back home on Cape York. I made this over many months of journeying along songlines all over the continent from east to west, visiting sites where the old Law is still kept in sacred places in the land, reflected in the stars, and in ceremonial objects and story held by our old people. On one side of the boomerang is the Seven Sisters constellation, and on the other is the Orion constellation, and the songlines I traveled are cut round and round the handle, intersecting with each other.

The stories of the old fellas who have shared their Law with me are in there. The boomerang is charred black from the fire I worked with in making it, but when you put it under water, the dark red ochers are activated and glow like the eyes of Oldman Juma's toads, tracing that Star Dreaming of the seven spirit families who are working to bring all our mobs back together again and begin a new epoch.

There are people today fighting to transform our Law into a hybrid entity that may be recognized in Australian and international law, and then there are the old fellas who keep the original Law for us, holding it against the day of resurgence that will come.

Those old fellas don't want to be written about or filmed. They just keep our Law in secret places and sacred objects that are so powerful they take your breath away. They have no need to assert or defend this Law. It is immutable and will outlast anything you can inscribe on paper or store on a server.

This Law cannot be extinguished by the weak curses of land-use agreements and native title policies. This Law cannot be changed through dialogue either—it is the authority that shapes and regulates dialogue to keep it within the sustainable patterns of creation. It is neither the irresistible force nor the immovable object. It is neither the action nor the reaction. It is the thing in between.

Be Like Your Place

There has been a drought for ten years, and us-two are out on stone country near Tibooburra in northwestern New South Wales with some Elders, singing a rain song. We're excited, but the old fellas are not because they can see the pattern and we're having trouble seeing beyond the deep-down thrill of power, the possibility whispered from our emu/frilly lizard egos, *I can control the weather. I am greater than.* So we miss the pattern and see only drought relief and rivers flowing strong again and a hell of a story to tell around the fire. We are children who don't know what else will come from this, and we don't give it a second thought. So we sing and we dance up a storm.

What the old fellas can see are the beetles already climbing the riverbank and the swifts flying urgently and close to the ground because they know what is coming, signaling all the other animals to move to high ground and plants to shift their energies into strengthening their roots. Gidjirr trees also signal a warning by emitting a powerful scent that smells like rotting broccoli. The long-suffering sheep and cattle and cotton crops don't catch these signals though, and even if they did they are incapable of moving with the land and weather or digging in. Their owners have fenced in their livestock and are watching meteorology reports on television, in houses built on floodplains.

The rivers will flow and then burst their banks and overrun the levees around small towns. Shops will close, and families will go hungry for a while. The rivers will flow again, but that water will be halted by dams at Menindee and released when the appropriate officers see fit. It will change things in multiple systems in three different states. It will be halted again by floodgates at the Murray River mouth and released into the sea months later, creating the dynamic mix of fresh and salt that the sacred system of the Coorong depends on but hasn't seen in years.

There will be a dust storm that will sweep across the state from Tibooburra and cover Sydney in red earth. Elders there will see this as a big ceremonial action marking the streets and buildings and people with the color of red ocher. People will have to replace their stained clothing, and the malls will experience a spike in trade. The red dirt will go into the sea and cause a massive algal bloom, clearing the air in a huge carbon sink while temporarily restoring fish stocks as they feast on this sudden abundance. Unemployed fishermen will enjoy a brief revival of their industry, and markets will fluctuate along with welfare demands that will fall and then rise again when the fish are decimated once more. Inland, settlers who had abandoned their dusty properties and left small towns in droves will return with the green grass, and the Aboriginal residents will be in the minority again. The activities of the police force will ramp up, along with rents and housing prices.

It takes a long time before the water reaches the mouth of the Murray River, long enough for us-two to travel down to the Coorong to be present for the moment when the flood-gates are opened. Dead penguins and seal carcasses have washed up on the beach, and the place feels postapocalyptic as we stand on the sand and wait for the fresh water to come.

We see it tumbling toward us, swirling and frothing, while, out to sea, shapes are moving in the deep. They spiral and play there frantically, and one of them breaks away and swims toward us. He arrives at the shore just as the fresh water hits in the same place, and he lumbers across the sand to sit in front of us.

This is not like any seal we've seen before. He has a beard and is barking in a human language that our brains don't recognize but that resonates somewhere deep in our bones. He glares at us with his fathomless eyes, then turns and shuffles back into the sea.

Everything is creation, and there are always patterns to perceive. If wombats are on the move, the sap is running in the gum trees and it's time to cut bark. If the tea trees are flowering, lychees and cherries will be available at the supermarket, where "Jingle Bell Rock" will be playing in an interminable loop. Controls preventing capital flight are announced in one country, and in another country the real estate market will plummet and interest rates will come down in response to rising unemployment. Loans will be taken out to fund infrastructure projects that will buoy up the construction industry, the biggest employer. Later, cuts will be made to social programs when those loans need to be repaid. And then there is the weather.

In the US, a couple hires a weather modification company to prevent it from raining on their wedding day, and there is a forest fire in their area two weeks later. Elsewhere, iron filings are dumped in the ocean to create algal blooms for carbon capture experiments in climate engineering. Silver oxide is sprayed in the sky to seed clouds for rain, temporarily clearing the pollution from a city that is hosting a sporting event. Thailand innovates a cloud-seeding technique that makes it a world leader in the field, while online companies advertise their weather control expertise to governments all over the world.

I see on the website for Weather Modification, Incorporated, that an Australian state government agency has procured equipment, pilots, and training for an extensive cloud-seeding project involving the spraying of toxic elements in the sky. I write to the agency asking for any research they have done on the environmental and human health effects of this program, and they respond with links to web pages showing research into the levels of rainfall produced by the project. I ask again for health and environmental impact studies and receive no reply. Weeks later I return to those web pages to find they have been removed.

If land and people are not even considered as variables in these weather experiments, then it is certain all the interrelated elements of dynamic weather systems and the knock-on effects of geo-engineering are not informing these activities either. The people conducting them are like children doing a rain dance. Future survival of all life on this planet will be dependent on humans being able to perceive and be custodians of the patterns of creation again, which in turn requires a completely different way of living in relation to the land.

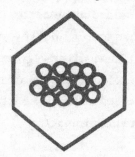

There is a pattern to creation. In this image you can see the way that pattern is expressed through Turtle story. A giant Turtle Spirit is hit with massive force at the center of his smooth shell. The impact makes a round section that cracks out to form another, and another. The interconnected pressure of all these round parts together forms them into hexagons, like in a sugar bag (honeybee hive). I couldn't believe it when Oldman Juma told me the sequence—one became two, then three, then five in a row along the center of the shell. Then, at the eight points where each part met, eight new parts formed. The sequence goes: 1, 2, 3, 5, 8, 13, and so on.

I laughed and said that's the Fibonacci sequence, "discovered" by an Italian mathematician around eight hundred years ago. Might add a couple of zeros to that date—it's been around longer than that. It's the pattern of creation, forming what some call the "golden ratio" that all nature is built upon, from flowers and trees to your body and even DNA.

With the impact at the center, all the pieces move outward but eventually move back inward again. This action includes the movement of people. In this story, people were created from the center of Australia, and many groups spread throughout the world from there. At this point in history, all those groups have returned to these shores and all the peoples are represented here again. The families have all come back together in this place, awaiting the next impact.

I carved this chapter into another killer boomerang. The killer boomerang is asymmetrical and far removed from the popular conception of the returning boomerang, which was traditionally used as a toy or in some places as a decoy imitating the flight of a hawk. But this boomerang goes straight and breaks legs. Similar flat, angular throwing sticks have been made for millennia by cultures all over the world. There was a crate full of them in Tutankhamen's tomb. An ancient boomerang made from mammoth ivory was found in Poland. But no culture in the world has produced the diversity and sophistication of Australia's boomerangs, of which there are thousands of different varieties. They emerged from cultures steeped in the patterns of creation and therefore are immensely adaptive and innovative in response to diverse and shifting landscapes. You have to be living the patterns of your place if you want to tap into this kind of genius.

The killer boomerang I carved is etched all over with patterns I perceive in the interaction of fresh water and salt water and the movements of sand along the coast. There are points where the pattern is disrupted by entities that are out of relation to the rest of the system. There are also points where the pattern reasserts itself through strange attractors, entities that adapt and spawn new responsive networks of agents in the self-organizing system.

In the home language of my family, there is no word for culture. There is a phrase that approximates that concept, but the meaning is untranslatable into English: *aak ngam-param yimanang wunan.* If you look at the direct translation of each word, the meaning comes out as "being like our place." I think this is a good way to start if you want to begin to discern the patterns of creation and rejoin our custodial species. When you engage with this way of being, you will find it changes you in subtle ways. The patterns of your language will change as you find ways to express the places you come into relation with. Your accent will change to reflect the landscapes you inhabit. I have moved around so much in Australia over the years that my own accent is weirdly muddled: there are too many places and Peoples in me. Being in profound relation to place changes everything about you— your voice, your smell, your walk, your morality.

But I can't explain how to do this, nor can I show you by taking you with me walking out in the bush and through all the layers of earth and water and sky. So we'll have to find a nonphysical way to do it, a deep visualization stimulated by speech-sound marks on a page.

Deep visualization is like a poor man's virtual-reality machine. But I've tried virtual reality and don't remember much about the content (something about shooting zombies). I do, however, remember everything I've experienced through powerful, deep imaginings. You close your eyes and relax, gradually coming into a deep state of ancestor-mind focus, while you go through a vivid imagining of a process or a story.

In this activity there are no limits to what you might experience—you can travel the solar system, walk through a market where vendors only speak Cantonese, learn how to drive a forklift or do a liver transplant. Use rhythmic language where possible, alliteration and rhyme with repetition—these are the devices oral cultures around the world have always used to assist in the transformative application of spoken texts. These devices survive mostly in poetry and song today but have lost their true place in complex knowledge transmission, with the false divisions that have arisen between arts and sciences.

The remainder of this chapter is a text I have used to bring many people to a profound understanding of being in place. It is a dream walk that I exhibited as a sound installation in an art exhibition called *Revealed*, in Melbourne in 2017. You might try reading it aloud to someone or to a group of people while they sit or lie down with their eyes closed. Afterward, we'll connect all the dots.

Here now. Picture in your mind a campfire you have seen, from a time when you have felt safe and happy and connected. See the fire, see it clear, have no fear. See deep into the heart of the hearth to the flame in the hot coals. See it dance, see the rhythm of it flickering. Feel the heat on your hands and face. Rub your hands together, make them hot, then rest them on your belly just under your navel. Rest them warm on your belly and feel the heat there.

Picture and feel the campfire in your mind. It is a real thing, this image. It exists, but where is it? Is it in your brain? No, it is in your mind. But where is your mind if not just in your brain? Imagine for a minute that your mind is not trapped only inside your skull, that it can move around your body. See the campfire in your head, crackling yellow and red. Feel the heat of it reach your face. Now try to move it down, that image; see it move and feel the warmth of it move, down your face and past your chin, down your neck and to your chest; see it and feel it move down past your liver, down over your stomach to your lower belly area.

Hold the image of the fire right there in your belly. It is not in your head now. Your mind is more than your brain, and your sight is more than what your eyes provide. Your mind's eye can see the fire there in your belly. See it, feel it burn. There is a power there, in your belly. This is your big power. It has energy, warmth, and a rhythm, just like the fire you are seeing there with your mind's eye. Feel it under your hands, burning bright. Now blow it to burn brighter.

Take a slow, deep breath. Blow it gently out through lips that are almost closed, the way you blow on hot coals to make them glow. See the coals crackle and spark. Your hands are protecting the fire in your belly. See it as you breathe in again slowly and blow right into the heart of the hearth fire. Breathe in, breathe out, see the coals glow and the flames flare. Keep breathing in and out now, fanning the flames. The air from this place, as it enters your lungs, is leaving small pieces of itself behind that enter your bloodstream and race through it to every part of your body. Those pieces are like sponges, mopping up tiny pockets of poison, poison left behind from a thousand sad feelings, bad memories, toxic events, and attacks and accidents and damaged lands, stolen places, stolen things, corrupted waters, murders and massacres, fences and blocks and assaults and insults and injuries and hidden histories. Let those poisons be mopped up by the little sponges of air in your blood, let them be taken in those sponges all through your body and back to your lungs, moving back from liquid to air and blown back out through your mouth and nose. Blow the badness out, let it leave you and disintegrate in the air, rise up far above, and blow away on the breeze.

The poisonous things leave behind dry areas, damaged areas, in your body. Imagine all through your limbs and torso and head, all this dry and crackly dead grass, snapped branches, and gray, papery leaves. All this scrubby trash is blocking your mind from moving freely around your body. See the fire there in your belly. Blow again on the hot coals, gently, seeing them glow and feeling the heat. Blow again, harder this time, and see the sparks begin to spiral up from the fire as you blow. Blow again, harder, and make the sparks fly further from the flames, fluttering upward into the dry, dead grasses laying thick under your ribs. See them land there, settle in the feathery fronds, smoking softly in wisps that wind up through the itchy snarls of grasses and sticks on layers of leaves. See the tiny flames flare and climb and spiral and spread, and a sparrow hawk swoops in, wings in the rhythm of the flames, to take one of the smoking sticks in its talons, sweeping off inland to drop it smoldering in the knotted undergrowth of your chest. The fire forms a long line burning across your torso, dancing the same rhythm as the fire in your belly, bright yellow flames that burn briefly and not too hot, leaving soft, cool ash that crumbles into the soil, your flesh, and nourishes it, and will continue to nourish it each day as the dew drops fall and settle and sift inward.

The line of fire spreads upward. All the scars and injuries and brittle patches left by damage and loss and abuse are being swept into the fire and made into soft charcoal and ash, swirling together like crow and white cockatoo feathers. Soon it will spread to your throat, then your eyes, then your ears, then your hands. See the flames and smoke and feel the heat and hear the crackling and snapping through your ribs, your collarbones, your neck and throat, mouth and chin, nose and cheeks, eyes, forehead and scalp, ears, shoulders, arms and elbows, forearms and hands, all burning, and the smoke billowing out through your fingers and high into the sky. Dry, damaged areas are all made clean and new and warm and light. The fire spreads now, from your hands, past the campfire hearth in your belly, flames leaping across from your fingers to your hips, burning through the dry grass there and on through your loins and buttocks and thighs. The fire has a rhythm—feel it dance in your legs. Your knees pop and crackle as they go up in flames, your shins and calves, your ankles and feet, until smoke billows out of your toes and high into the sky, leaving you light and clean and new. Dew drops fall and cool everything down, dampening the ashes and soot and soaking it into the ground where heated seeds sprout and begin to take root. Country is becoming well. You are Country. You are becoming well.

Think of somebody you love unconditionally, no matter what. If you can't do this, you need to stop now and go take care of your life. If you can, picture that person in your mind and let yourself feel that love for them. Feel it like a sensation coming over you. Keep seeing the image of that person and feeling that love, then picture them in a special place. That place will be somewhere outdoors that is special to you both, a place where you have shared joy and love together, where you have connected deeply with the land and with each other. See every part of the landscape around you, the plants, trees, dirt, or stones. See the person you love holding a big bucket of water. They tip it out onto the ground. They follow it where it flows, but let it ignore any civilized or synthetic barriers. Follow them, the loved one and the water, and pay attention to the way the water moves on the ground in that place, where it goes, where it stops, where it sinks. Feel the love for that person, and then stretch out with that feeling, and feel the same sensation for that place. Let your love move all across it, through it and into it, the same way rainwater would if it fell there right now. Feel that love spreading all over that Country. Where do you feel it in your body? In your chest, head, belly? Is that all? Is that feeling just inside you, or does it go further? Can you feel that love outside yourself, a long way off, in that special place? It is right there, a part of you and a part of the one you love, a feeling, a part of your mind and spirit, right there in that place. Can you feel it?

Holding on to that feeling, let the picture of your loved one and that place fade gently into the background. Slowly replace it with an image of the place where you are now. Not a room or a building—allow all that to be transparent. See the land, the ground, the waters and landforms around you and beneath you. Hold the feeling of love. Imagine where the water would flow if it fell on that ground right now. Let the love trickle out from you, lapping outward in ripples or tinkling in streams through the earth, see it there all around you for that place, feel it in the place. Feel the place. Love the place. There may be wounds or sickness there in the land that makes you sad, but hold that loving feeling because it is unconditional.

The love is not just a feeling in your body now, or in a distant place, or with another person. It is around you in the place where you sit, in the land, along with the feeling from all the old people who have been connecting the same way with that place for thousands of years. All the memories of those Ancestors are there. All your own Ancestors' memories lie inside you in the same way, in your bones, in every part of you, in your cellular memory. Mind and memory are real things although they can't be touched, measured, proven, or even seen. They exist, but not only in your brain. They extend out, to your body, to the land and your relations. Your mind is infinite and extends as far as your attention and love can go.

Mind, brain, body, land, loved ones—all these things at the front of your thoughts at once; they make you heavy. Your body and spirit sinking, thinking, heaviness moving deep through you. Eyelids heavy, closed; hands heavy on your belly, back pressed into the ground—skull, shoulders, feet, legs, all heavy. The feeling, the loving in the land and the love in your body are the same thing, you and the earth the same thing now, so you sink right into it. Feel yourself going through the floor or ground like it's quicksand, but transparent, the land, you, sinking slowly into the dirt or sand beneath you. There are layers of rock and water down here, and you pass through them, sifted and cleaned by them as they pass through you, cooling you as you fall deeper and deeper into the dark. You are not afraid. You are not alone. You begin to feel that loving feeling again, only now it is not coming from you, but coming into you, flowing through you and around you. There is a warmth and a rhythm to it. A deep rhythm sounding deep, deep down. You sink further, faster toward the loving rhythm and the warmth.

You see a soft glow, closer and closer, warmer and warmer, the rhythm thicker, thicker and older and stronger and deeper than anything else. It is familiar, comforting, and so is the glow. The glow is from a fire, a hearth fire or a heart-fire with hot coals. You know it. You've seen it before. Could it be the sun? It is the same as the fire in your belly, the image in your mind from before, only it is massive. Its power is made of the same stuff as the power you carry in your belly, your big power. As you feel the heat and rhythm, you sense the same inside you, beneath your hands, thrumming and glowing. It feels like a fish sniffing at a line, deep beneath the sea. It is faint but real at your fingertips, stirring deep in your belly. You know this. You remember this place, the earth you came from, the big mother that bore you. You're home. Now that big hearth fire under the earth stokes your own fire, the fire in your belly, replenishing your power, filling you up with endless love and energy. It is clean, warm, and pure, and it lifts you up. You feel lighter, and upward you go.

The glow disappears beneath you, but the rhythm and flame remain in your belly. They propel you up, up, up, up through miles of heat and rock and cold and earth and more rock and sand and water and earth. Up, up, up until you reach the place where your body is sitting or lying there in the world. But you can't stop. You keep drifting upward now, lighter than burnt leaves on the breeze, stirring and rising up, up, up through any ceilings or tree tops, up, up through any clouds and into a shining blueness that stretches as far as you can see all around. Up and up still you rise, as the blue becomes deeper, richer, darker. Soon it falls away beneath you, and you are into a clear black infinity all around. The earth is far below, and you drift on through burning stars.

The Seven Sisters are here, burning bright. All the hero Ancestors are up here, sky camp, watching you, blazing, that same fire again. Your rhythm is pulsing out to them; they are pulsing back, light washing through you in waves, stoking up that fire inside you again, blazing those coals, washing you clean and clear. Can your mind even extend up here, to the patterns in the endless night sky? Can your mind possibly perceive all of these stars, shapes, gaps—forms made by those gaps, the stories and morality and rich knowledge here, thousands of parts, all at once? Can it know every part at once? What it does hold for sure are the patterns created by all these parts. It sees objects pulled toward a space and knows what must lie there out of sight. It is able to see the swirl and stories and positions and angles and times and seasons all at once, to read the big patterns these show together and therefore make predictions and judgments about land-based events, phenomena, weather, ritual, the timing of all things in your life throughout each cycle of seasons, the longer cycles of generations, and the even longer cycles of deep time and story.

What would it take to free your mind, to allow it to see these big patterns again? All the Ancestors up here, they left their traces in the earth and waters below as well, and you carry those traces within, those memories and knowledges and deep, deep love. Those things wait for you, below. They tug at you, begin to draw you down. You are no longer light, but not heavy either. You are in balance, and you return to your place of love below. You are a point of connection between the earth and sky camp, so go, be that. You drop, plummet through stars and darkness and blue-black and deep blue and light blue and maybe clouds and water drops and tree tops and ceilings and then softly, softly, softly settle back down, down to earth, into the feeling of your place and your body, the rhythm that never stops, the fire in your belly, your power, and the infinite potential of your mind, within and without.

Which Way

In 2008 a thousand thinkers from all over Australia descended on Parliament House in Canberra for Prime Minister Kevin Rudd's 2020 summit, to find the "big ideas" to take the country into the next decade. I was one of the delegates, and each of us had to bring along an idea that would be considered and potentially adopted by the government. My idea was to create Indigenous Knowledge centers all around Australia where Aboriginal and Torres Strait Islander people could gather thinkers and knowledge-keepers together and grapple with solutions to the world's sustainability issues.

To my surprise and great joy, my big idea was taken up and funded with the full support of the government. A process of consultation was undertaken in communities all over the country to design these centers of knowledge and sustainability. Can you imagine it? First Peoples networking and researching and innovating, using the same kinds of ancient knowledge processes I have described in this book.

Our Knowledge would take a respected position, producing amazing solutions to the many crises afflicting the world. Our Law would become known and respected as the most sustainable basis for living and being on this continent. The Australian people would be able to move on from "sorry" and start saying "thank you" (and maybe, after a while, "please"). It was going to change everything! An entirely new industry was about to be created that had the potential to lift our communities out of poverty and into a position of dignity.

After the consultations, I didn't hear anything about it for a long time, and a couple of years later I visited one of these knowledge centers in a capital city. It was not exactly what I had hoped it would be. There were a couple of staff there, curating exhibits in a beautiful room filled with artifacts in glass boxes, dot paintings, and stirring tributes to Indigenous Australia's most famous sporting heroes and country music artists. And free Wi-Fi.

I'm quite sure that this was the moment when I stopped being a strange attractor in a complex dynamic system.

There is no carving for this chapter, because the mnemonic device is my hand. Your hand. Us-both hand. Yup, it's literally in our hands. We can put that image anywhere, printed in the sand or in ocher on a rock or tree or object, in more numerous copies than there could ever be of this book. Here's how we can resolve the mass distribution limitations of having Indigenous Knowledge coded into carved objects—most people have hands, so if we put all the knowledge into them, anyone can carry it around. I considered 3-D printing the carvings for each chapter and providing that code with the book so everybody could hold them, but that would just be lazy and unsustainably resource-intensive. So we're going to work with our hands.

I have met a lot of Elders who encode knowledge in their hands as a kind of memory aid. Oldman Juma has stories buried in every crease and knuckle. Noel Nannup in Perth has sites and creation Ancestors in each finger so that he can stand in one place and stretch out his hand with each digit pointing in the direction of five different sacred sites in the landscape. He taught me how to touch them together in different combinations and find new connections between knowledges and places.

Hands are part of my own totemic system, and they are a sacred symbol all over Australia. You may have seen handprints and stencils on rock art here, and even in cave paintings from around the world. In all the different Aboriginal languages, this word—"hand"—is one that often sounds the same—usually starting with the sound *ma*. Strangely, this is also the start of the Latin word for hand, *manus*. Oldman Juma says, "See?" The hand is a symbol that is sacred to every culture, a meaning remembered everywhere; just like the Seven Sisters constellation that carries the same name across the globe—a remnant of a common origin and a call to come home.

It is the cultural lens that we carry everywhere with us. Remember when I showed you a hand gesture demonstrating that perspective, carrying the message that Indigenous Knowledge is not about the what, but the how? It is about process, not content. Your culture is not what your hands touch or make—it's what moves your hands.

I recently published a research paper sharing five ways of coming to knowledge that are present in the Wik Mungkan language. The first is *mee'-aathan*, which means learning through close observation and demonstration. The second is *ma'-aathan*, which means passing on knowledge with a helping hand, assisting somebody, then gradually stepping back. The third is *thaa'-aathan*, which means passing on knowledge verbally in activities like yarning. The fourth is *konangam pi'-pi'an*, which is memorizing through deep listening. The fifth is *ngantam ngeeyan*, which is thinking, reflecting, and understanding.

I remember all of these by encoding an image on each digit, starting with the pinkie—an eye, a hand, a mouth, a left ear, and a right ear. The thumb, which is for thinking and reflecting, is touched to each finger as I go through and recall them. Every Indigenous culture has its own unique style for transmitting, producing, and storing knowledge, and this is the one I am most familiar with. There are common orientations across the continent though, and these are the ones I have shared with you in this book.

You might recall them from earlier: The child in your

pinkie for kinship-mind. Its mother in the ring finger for story-mind. Her husband in the middle finger for dreaming-mind. His nephew in the index finger for ancestor-mind. All of them connected in spirit through pattern-mind in the thumb. Notice how each metaphor used as an image is connected in pairs and groups to the other images. The information is in each part, but the knowledge lies in the connections between them.

We could apply the same technique to remembering some of the other key ideas we have been yarning about in this book. The most important one is the set of protocols for agents in a sustainable complex system. If you are interested in being this kind of agent, we can encode that in your fingers right now. Not in the first row of knuckles near the tips of your fingers, though, because that family lives there with those five different ways of thinking. The complexity agent protocols will go on the middle row of knuckles. When we're done, you'll know this like the back of your hand.

First you need to come up with an example of a dynamic, self-organizing system that you know well: a solar system or river or family or waterway or ecosystem or football team—anything like that. It has to be something that carries particular meaning, stories, and connections for you. See an image of that system in your mind, and then stare at your thumb, thinking the image into it.

Now choose four elements from that system to represent the protocols of a complexity agent—connectedness, diversity, interaction, and adaptation. You must form an image of each element that connects to what you are trying to memorize. Your thumb will touch the inside of each finger

at the second knuckle as you think the image into the finger.
It may help to break each concept into three parts and associate those with different parts of each image. For example,
the three parts of the connectedness protocol are pairs, networks of pairs, and networks of networks.

Connectedness involves forming pairs with multiple other
agents who also pair with others. The next step is creating or
expanding networks of these connections. The final step is
making sure these networks are interacting with the networks
of other agents, both within your system and in others.

Recall that the diversity protocol has three parts—
similar to you, different from you, and systems beyond you.
It compels you to maintain your individual differences, particularly from those other agents that are similar to you. You
must also seek out and interact with a wide variety of agents
that are completely dissimilar to you. Finally, you must interact with other systems beyond your own, keeping your
system open and therefore sustainable.

The interaction protocol is about continuously transferring knowledge, energy, and resources. This means passing
on these three things to as many other agents as possible,
rather than trying to store them individually.

The adaptation protocol is about transformation, feedback loops, and strange attractors. You must allow yourself
to be transformed through your interactions with other
agents. Knowledge and energy will flow through the entire system in feedback loops, and you must be prepared to
change so that those feedback loops are not blocked. If you
are truly adaptive and changing, you are open to sudden
eruptions of transformation, in which you may temporar-

ily take on the role of strange attractor and facilitate chain reactions of creative events within the system.

On your fingers now you have two familiar sets of ideas from this book so far. There are some ways of thinking about the world and some protocols for sustainability, but so far no process for enacting these things. So now I will share with you a cultural process for enacting these ways and protocols if you wish. I will leave it up to you to figure out how you want to store these ideas on your last row of knuckles.

This process for working with Indigenous Knowledge in sustainable ways came from two years of following song-lines from north to south and then east coast to west coast with Mumma Doris Shillingsworth. We wanted to find a pattern that was common to all different groups, a process that could be used by anyone to come to Indigenous Knowledge productively and without doing damage. We had hundreds of yarns, and Mumma Doris fairly talked my ear off in the car, so much that once I nearly crashed trying to take it all in. She worked with her knowledge of Moondagutta, which is the Murrawarri word for the creation snake, linking up that songline all the way to Waagal (Rainbow Serpent) story of the Nyoongar in Perth. Every yarn on the way was first grounded in the protocol of respect, which every Elder insisted on, so that became our touchstone. We had four questions:

1. What can we know?
2. What do we know?
3. How do we know it?
4. How do we work with that?

The answers we found were as follows:

What we can know is determined by our obligations and relationships to people, Ancestors, land, Law, and creation.

What we know is that the role of custodial species is to sustain creation, which is formed from complexity and connectedness.

The way we know this is through our cultural metaphors.

The way we work with this knowledge is by positioning, sharing, and adapting our cultural metaphors.

What we had found was a broad, common description of Indigenous ways of valuing, ways of being, ways of knowing, and ways of doing. These things had a widespread order, a sequence in all cultural activities in which people were sharing or producing knowledge on Country. We had our own personal metaphors for describing this process of induction. I referred to it as spirit, heart, head, and hands. Mumma Doris knew it as Respect, Connect, Reflect, Direct. She insisted on that order. She also noted that non-Aboriginal people seemed to work through the same steps but in reverse.

Mumma Doris has observed interventions and programs imposed on her community for over half a century, noticing that they always begin with the last step, Direct. Government agents come into the community with a plan for change, and they direct activities toward this change immediately. When it all fails, they go backward to the next step, Reflect. They gather data and measure outcomes and try to figure out what went wrong. Then they realize they didn't form relationships with the community, so belatedly they go to the next step, Connect. Through these relationships they discover the final step (which should have been the first), finding a profound respect for members of the community they ruined. They cry as they say farewell and return to the city, calling, "Thank you! I have learned so much from you!"

Invert that process and you'll have something approximating an appropriate way of coming to Indigenous Knowledge and working toward sustainable solutions. The first step of Respect is aligned with values and protocols of introduction, setting rules and boundaries. This is the work of your spirit, your gut. The second step, Connect, is about establishing strong relationships and routines of exchange that are equal for all involved. Your way of being is your way of relating, because all things only exist in relationship to other things. This is the work of your heart. The third step, Reflect, is about thinking as part of the group and collectively establishing a shared body of knowledge to inform what you will do. This is the work of the head. The final step, Direct, is about acting on that shared knowledge in ways that are negotiated by all. This is the work of the hands.

Respect, Connect, Reflect, Direct—in that order. Everything in creation is sentient and carries knowledge, therefore everything is deserving of our respect. Even narcissists.

Echidna is the villain in Turtle story—so in the end, Turtle eats him. Today you'll find in the skeletons of freshwater turtles a big spiky ball of bones that serves no evolutionary or biological purpose, and that's Echidna in there still. It's a living reminder in the pattern of creation of those ancestral spirits who made the big bang and also a living warning to all narcissists who like to think they are greater than. But Echidna is only the villain during the events of the story. After he is punished, he doesn't hold on to that guilt, and others don't hold that blame for him. They let it go, and he lets it go, learns from his mistakes, and becomes something wonderful—the species with the largest brain in the world in relation to his body size. Echidna is often somebody who does terrible things in our stories—why else would he end up with so many spears thrown into him? Turtle does not hate him, because without that struggle the universe wouldn't even be here. Echidna doesn't hate himself either, for the same reason. Guilt is like any other energy: you can't accumulate it or keep it because it makes you sick and disrupts the system you live in—you have to let it go. Face the truth, make amends, and let it go.

My hope is that one day everybody can find a place under the Law of the land where they live, transitioning our living systems into something that is sustainable in the true sense of the word. Oldman Juma calls it the seven families coming home and uniting again. We'd like everybody to look up at the stars and see the same stories there once more. And stop asking the question, "Are we alone?" Of course we're not! Everything in the universe is alive and full of knowledge.

Now that the small questions of existence have been answered—*Why are we here? How should we live? What will happen when we die?*—us-two should be able to get back to the business of asking some of the bigger questions. We'll need living lands and bodies to do that, though. So let's put these hands of ours to work.

Acknowledgments

I am grateful to the following people for yarning with me, sharing knowledge, and giving me permission to work with their ideas, words, or stories: Michael "Ghillar" Anderson, Kerry Arabena, Vincent Backhaus, Michelle Bishop, Murray Butcher, Aunty Beryl Carmichael, Jacob Cassady, Kristy Chua, Aunty Min Collard, Jayson Cooper, Aunty Gail Dawson, Max "Targaryen" Dyers, Juma Fejo, Ty Gordon, Dr. Larry Gross, Mick Harding, Paul Hughes, Hayden Kelleher, Megan Kelleher, Melissa Kirby, Aunty Janice Koolmatrie, Priscilla and Sarah Loynes, Clancy McKellar, Greg McKellar, Kelly Menzel, Donna Moodie, Dr. Noel Nannup, Robyn Ober, Luke Patterson, Percy Paul, Ellie Rennie, Adam Ridgeway, Aunty Jenny Robinson, Aunty Doris Shillingsworth, Lauren "Blackie" Tynan, Kakkib Li'Dthia Warrawee'a, Tiffany Winters, John Zerzan.

And I am grateful to the following writers whose books helped me with my thinking:

Aristotle, *Corpus Aristotelicum*

Noam Chomsky, *Power Systems: Conversations on Global Democratic Uprisings and the New Challenges to U.S. Empire*

Dante Alighieri, *The Divine Comedy*

Barbara Ehrenreich, *Smile or Die: How Positive Thinking Fooled America and the World*

Nicholas Evans, *Dying Words: Endangered Languages and What They Have to Tell Us*

Susan Faludi, *Backlash: The Undeclared War Against American Women*

Keith Farnish, *Underminers: A Guide to Subverting the Machine*

Bruce R. Fenton, *The Forgotten Exodus: The Into Africa Theory of Human Evolution*

Muammar Gaddafi, *The Green Book*

Edward Gibbon, *The History of the Decline and Fall of the Roman Empire*

Germaine Greer, *On Rage*

Edward T. Hall, *Beyond Culture*

Stephen Hawking and Leonard Mlodinow, *The Grand Design*

Paul Keating, *Engagement: Australia Faces the Asia-Pacific*

Aileen Moreton-Robinson, *Talkin' Up to the White Woman: Indigenous Women and Feminism*

Martin Nakata, *Disciplining the Savages: Savaging the Disciplines*

George Orwell, *Homage to Catalonia*

Bruce Pascoe, *Dark Emu: Aboriginal Australia and the Birth of Agriculture*

Thomas Piketty, *Capital in the Twenty-First Century*

John Pilger, *A Secret Country: The Hidden Australia*

Plato, *Phaedrus*

All the books of Henry Reynolds

James Rickards, *The Death of Money: The Coming Collapse of the International Monetary System*

Carne Ross, *The Leaderless Revolution: How Ordinary People Will Take Power and Change Politics in the Twenty-First Century*

Edward W. Said, *Orientalism*

Linda Tuhiwai Smith, *Decolonizing Methodologies: Research and Indigenous Peoples*

All the novels of Sheri S. Tepper

Kakkib Li'Dthia Warrawee'a, *There Once Was a Tree Called Deru*

Alexis Wright, *The Swan Book*

About the Author

Tyson Yunkaporta belongs to the Apalech clan from Western Cape York and is a senior lecturer in Indigenous Knowledges at Deakin University. He has worked extensively with Aboriginal languages and in Indigenous education, and his research activities include oral histories of natural disasters, language, health, and cognition. He is a published poet and exhibited artist who practices traditional wood carving.